Accident Investigation Training Manual

Accident Investigation Training Manual

Arnold G. Wheat

*Accident Reconstruction Specialist,
ACTAR Reg. No. 226*

Australia • Canada • Mexico • Singapore • Spain • United Kingdom • United States

Accident Investigation Training Manual

Arnold G. Wheat

Vice President, Technology and Trades SBU
Alar Elken

Editorial Director, Professional Business Unit
Gregory L. Clayton

Product Development Manager
Kristen Davis

Production Editor
Benjamin Gleeksman

Editorial Assistant
Vanessa Carlson

Cover Design
David Arsenault

Marketing Director
Beth A. Lutz

Marketing Coordinator
Marissa Maiella

Production Director
Mary Ellen Black

Procution Manager
Larry Main

Production Services
TIPS Technical Publishing, Inc.

COPYRIGHT © 2005 Thomson Delmar Learning, a part of the Thomson Corporation. Thomson Delmar Learning, Thomson, and the Thomson logo are trademarks used herein under license.

Printed in Canada
1 2 3 4 5 XXX 08 07 06 05 04

For more information contact
Thomson Delmar Learning
Executive Woods
5 Maxwell Drive, PO Box 8007
Clifton Park, NY 12065-8007

Or find us on the World Wide Web at
http://delmar.com

All rights reserved. No part of this work covered by the copyright hereon may be reproduced or used in any form or by any means—graphic, electronic, or mechanical, including photocopying, recording, taping, Web distribution or information storage and retrieval systems—without the written permission of the publisher.

For request permission to use material from this text contact us by:
Tel 1-800-730-2214
Fax 1-800-730-2215
www.thomsonrights.com

Library of Congress Cataloging-in-Publication Data

Wheat, Arnold.
  Accident investigation training manual / by Arnold Wheat.
    p. cm.
  Includes bibliographical references and index.
  ISBN 1-4018-6939-4 (alk. paper)
  1. Traffic accident investigation—United States. 2. Traffic accidents—United States. 3. Automobile drivers—Accidents—United States. 4. Evidence, Expert—United States. I. Title.
  HV8079.55.W49 2005
  363.12'565—dc22

2004023560

Notice To The Reader

Publisher and author do not warrant or guarantee any of the products described herein or perform any independent analysis in connection with any of the product information contained herein. Publisher and author do not assume and expressly disclaim any obligation to obtain and include information other than that provided to it by the manufacturer.

The reader is expressly warned to consider and adopt all safety precautions that might be indicated by the activities herein and to avoid all potential hazards. By following the instructions contained herein, the reader willingly assumes all risks in connection with such instructions.

Publisher and author make no representation or warranties of any kind, including but not limited to, the warranties of fitness for particular purpose or merchantability, nor are any such representations implied with respect to the material set forth herein, and the publisher takes no responsibility with respect to such material. Publisher and author shall not be liable for any special, consequential, or exemplary damages resulting, in whole or part, from the readers' use of, or reliance upon, this material.

Contents

Foreword ix

Acknowledgments xiii

1 Introduction and Definitions 1
General Definitions 2
State Definitions 3
Federal Definitions 4

2 Planning and Preparation 9
Corporate Policies and Procedures 10
Corporate "In-House" Resources 12
    Administration Department 12
    Human Resources Department 12
    Maintenance Department 12
    Dispatch Department 13
    Legal Department 13
    Driver 14
    Safety Department 14
Outside Resources 14
    Traffic Accident Reconstructionist 15
    Insurance Claims Adjusters 15
    Load Clean-up Services 15
    Hazardous Materials Clean-up and Remediation Services 16
    Towing Services 16
    Maintenance and Body Repair Facilities 17
    Legal Representatives 17
    Leased Equipment and Owner-operators 18
    Medical Facilities 18

3 Tools, Equipment, and Supplies 19
Photographic Equipment 20
Recording Devices 20

Measuring Tools   20
Reference Resources   21
Safety and Protective Devices   21
Miscellaneous Supplies   21

4   Law Enforcement Relations 25
Accident—Legal Definition   26
Law Enforcement Investigation   26
Traffic Accident Report   30
Working with Police at the Scene of an Accident   35

5   Investigative Factors 39
The Human Element   40
The Vehicle Element   40
The Environment Element   41
Time Frames   41
Elements of a Traffic Crash   43
   Possible Perception   43
   Perception, Reaction, and Response   44
   Encroachment   45
   Start of Evasive Action   46
   First Harmful Event   48
   Initial Contact and Maximum Engagement   49
   Disengagement or Separation   49
   Final Rest Position   52

6   Recording Techniques 55
Field Sketching   56
   Materials Needed to Prepare a Field Sketch   58
   Items That Might be Included on the Sketch   58
Coordinate Measuring System   62
   Establish a Reference Point and Reference Lines   63
Triangulation Measurement System   70
   Triangulation Method of Measuring a Curved Road   72
   Marking Materials   72
Accident Templates   76

7   Photographic Techniques 79
Vehicle Examinations   79
Highway Photographs   87

8   Vehicle Information 93
Damage Sustained in the Traffic Accident   101

9   Evaluation of the Highway Environment 109
Highway Information   110
Tire Mark Evaluation   123

10 Interviewing 141
Keys to Interview Success   142
Possible Areas of Inquiry During Post-Accident Interviews   144

A Resources 149
Organizations and Databases   149
Classes   150
Publications   150

Index   151

Foreword

This is a training manual and resource that I hope you never have to use. If this manual is consulted or utilized, then it suggests that some traffic event has occurred that was probably the result of a failure in the safety systems and driver training programs of a transportation organization, fleet operator, or motor carrier company. Its use may be the result of a driving error committed by another user of the highway system. Some type of vehicular or personal property was damaged and passengers were delayed. Freight and commodities were delayed, damaged, or destroyed during their shipment. At least one motor vehicle has sustained property damage. A driver, a passenger, or a casual bystander was either injured or killed. A tragic or catastrophic traffic event has occurred. A driver's confidence has been affected, and a professional driver's safety record has been blemished.

Traffic accident statistics developed by the National Highway Traffic Safety Administration and incorporated in a 2003 report titled, "An Analysis of Fatal Large Truck Crashes" identified several important facts relating to the involvement of commercial motor vehicles in traffic accident situations. They include the following:

- During the five-year period that the traffic records were analyzed, 18% of all large truck crashes were single-vehicle accident events. 62% of single-vehicle accident events during that same time period involved a vehicle type other than a large truck, indicating that this type of crash is more likely to involve a vehicle other than a large truck.
- Approximately 64% of all the fatalities in single-vehicle truck crashes involved a rollover movement by the large truck.
- Approximately 12–13% of all fatal traffic crashes were the result of a crash that involved a large truck.
- Approximately 80% of single-vehicle, large truck accidents occurred at locations not related to a driveway access or an intersection.
- The potential for rollover of a large combination truck increases 37% when the truck is operated during adverse weather conditions.

- When a large truck is involved in a crash event with another vehicle, the front of the truck is the initial contact area in approximately 66% of the crash events. About one half of those crashes involve the front of another vehicle striking the front of the large truck.
- Fatal traffic crashes involving large trucks are more likely to occur in highly populated regions, such as New Jersey, California, New York, Maryland, Florida, and Delaware.

These statistics, as well as others contained in the NHTSA report and other research reports relating to studies of traffic crashes involving large commercial motor vehicles, indicate there are difficulties and hazards associated with the operation of large trucks in certain highway configurations, traffic situations, and weather conditions. The ability of a trucking company to minimize the risks associated with the operation of a commercial motor vehicle often is a determining factor in the success of that trucking company. But to minimize the risk, company officials need to have an understanding of *how* and *why* their fleet vehicles are involved in traffic accidents. They need to know and understand what went wrong in their operation that allowed these costly and potentially tragic events to occur.

Traffic accident investigation is typically one of those areas that falls under the "umbrella" of responsibilities for personnel in the safety departments of a motor carrier corporation; yet it is often one of the subject areas in which the motor carrier safety professional has the least amount of formal training. Specific skills and technical knowledge related to traffic accident investigation often are not acquired or developed until there is a specific need for them. And although the success of a good traffic investigator often includes the application of common sense, a great deal of real-world, day-to-day experience, and some vehicle-related knowledge, the basic skills and specialized knowledge required to investigate traffic accidents are often lacking.

Therefore, when you receive notification that a traffic mishap has occurred, you must attempt to identify, develop, collect, record, and/or evaluate:

- Information, data, and physical evidence that you know will be present at the accident scene
- Damaged vehicles involved in the crash event
- Reports created by law enforcement agencies, ambulance companies, and fire rescue organizations
- Reports and assessments of insurance adjusters
- Reports of newspapers and news media organizations
- Observations and recollections of the various witnesses, drivers, and other persons associated with the event

The task can be a daunting one, because you realize that the manner in which you effectively identify, collect, and sift through the information will have a direct relationship to the outcome and value of the investigation. The continuation of employment of a company driver or the settlement of a claim against the trucking company may be dependent upon your professional performance and the accuracy and thoroughness of your investigative results from the traffic accident event.

In many circumstances, the lack of knowledge of an employee in the traffic or operational safety field regarding *how* to investigate a traffic accident may "slip through the cracks." The supervisor, accident review committee, or corporate official reviewing the information that you have developed and collected may possess a similar lack of knowledge about the subject of traffic accidents as you do. Unfortunately, the results of the investigation are accepted as correct and accurate, only because no one in the transportation organization, fleet operation, or motor carrier company knows any better.

A lack of expertise in traffic accident investigation is often magnified by the fact that, in a majority of transportation companies, the need to utilize these skills and this specialized knowledge is infrequent and minimal. Fortunately, most trucking fleet operators, transportation organizations, and motor carrier companies do not have a high traffic accident rate, even when operating large numbers of vehicles on the highways across the nation.

A recent research study by the University of Maryland's R. H. Smith School for Business, at the direction of the Federal Motor Carrier Safety Administration, identified several operational characteristics of successful trucking companies. Titled, "Highway Safety Practices: A Survey about Safety Management Practices among the Safest Motor Carriers," the federal-government funded study was based on anonymous responses to a lengthy survey provided to small, medium, and large motor carriers. The study identified the following characteristics and operational conditions for successful carriers:

- Hiring practices that identify safe drivers, whether they are company or leased drivers
- Written hiring policies that identify safe driving practices and attitudes
- Open and continuing communication between management and drivers
- Supervision programs and practices that identify unsafe driver characteristics and driving behaviors
- Continuous and in-depth driver training programs during employment
- Close monitoring of vehicle maintenance activities to insure safe vehicles
- Commitment to safety and safe operation above cost decisions

The difficulties in operating a safe transportation organization, fleet operation, or commercial motor carrier company are many. When a traffic crash occurs, there can be many questions about the reason(s) the crash occurred. Without a detailed and thorough investigation of the accident facts and an analysis of the traffic accident event, there can be no reliable answers as to why the accident occurred and how future incidents can be prevented or minimized. Successful transportation corporations, fleet operators, and commercial motor carrier companies have procedures in place to allow them to minimize the risk of traffic accidents, and they are able to identify and collect data in a timely manner. This process of identifying and analyzing sufficient data from an accident allows them to understand the problem when a traffic accident occurs and gives them the ability to implement changes in their system to reduce the number of future traffic collision occurrences.

Acknowledgments

When preparing a text such as this one, you realize how much you have relied upon the information, support and encouragement of others. I have been blessed by many such people. David Lohf and I have spent countless hours since 1982, debating observations and conclusions relating to the thousands of traffic accidents that we have analyzed together. His insight and knowledge have been amazing. James Garinger, a professional truck driver with several million accident-free miles during a 40 year career, continues to provide me background and expertise from the truck driver's perspective and fostered my interest in trucks. My office manager, Suzan Myers, has been a strong partner, providing continuity, leadership and excellence in our business for over 20 years. Finally, my wife and two daughters have tolerated my many absences and diversions, while providing love, support and encouragement at every step of my career.

Thanks also to the folks at Delmar Learning and to the production crew at TIPS Technical Publishing, Inc., including Robert Kern, Khedron de Leon, Juanita Covert, Ariel Tuplano, Meghan Greene, and Rachna Batra.

1

Introduction and Definitions

Current classification protocols of federal agencies that are responsible for analyzing and preventing traffic incidents and crash events—specifically the National Highway Traffic Safety Administration, the Federal Motor Carrier Safety Administration, the Federal Highway Administration, and the U. S. Department of Transportation—tend to not use the word "accident" when describing these highway collision events. The current trend suggests that an accident is not an "unintended event," but rather the result of one or more failures, incorrect maneuvers, or inappropriate actions made by one or more vehicle drivers that create a situation that results in a crash. The word "crash," however, has no technical definition within their literature. The crash category may include traffic events involving motor vehicles that are both intentional and unintentional, but which cause property damage, personal injury, or death.

In this text, the words "accident," "crash," and "collision" have been used interchangeably and refer to an unintended event, creating personal injury, death, and/or property damage, that results from the movement of one or more vehicles, as illustrated in Figure 1–1.

Although a review of definitions in a text is often boring and tedious, definitions relating to traffic accident events can have an important affect on how you, as the safety professional of a fleet operation, commercial motor carrier, or transportation organization, approach your job when an accident occurs.

Figure 1–1   A crash, accident, or collision scene. *Photo courtesy of Colorado State Patrol.*

General Definitions

The most basic definition of the word "accident" is one adopted by the National Safety Council, which states, "Accident is an occurrence in a sequence of events that produces unintended injury, death, or property damage." The National Safety Council further refined that definition by their adoption of the phrase "motor vehicle accident." That definition is contained in the *Manual on Classification of Motor Vehicle Traffic Accidents, ANSI D16.2-1996*, published by the National Safety Council. It states:

> Motor vehicle accident is an unstabilized situation that includes at least one harmful event (injury or property damage) involving a motor vehicle in transport (in motion, in readiness for motion, or on a roadway, but not parked in a designated parking area) that does not result from discharge of a firearm or explosive device and does not result directly from a cataclysm.

The definition of a motor vehicle accident is a United States standard by which all traffic accidents events or incidents are classified and which individual state traffic authorities use to determine the events included in their reports to federal traffic safety authorities. The definition is also utilized by many traffic safety organizations. The definition emanates from the *Uniform Vehicle Code*, which was first adopted in 1926 and was eventually adopted by all states to allow uniformity in traffic laws. The *Uniform Vehicle Code*, as well as the Model Traffic Ordinance, are updated periodically by the National Committee on Uniform Traffic Laws and Ordinances based in Evanston, Illinois.

State Definitions

A review of numerous state traffic laws indicates, however, that the word "accident," "traffic accident," "motor vehicle accident," "traffic crash," and motor vehicle "collision" are often not defined specifically in their respective state statutes. This is perplexing, as the statutes refer to the requirements and responsibilities of a vehicle driver regarding notifying and reporting to authorities an accident, stopping and rendering aid to the injured at an accident, and the necessity of a driver staying at the scene of an accident. However, the word "accident" is not defined within their respective statutes or ordinances.

In Colorado, for example, Title 42 of the Colorado Revised Statutes (which details specific statutes relating to vehicular traffic) does not define the word "accident" or any other similar words or phrases, such as "traffic collision" or "crash." A sampling of other states, such as Wyoming, California, Florida, and Montana, indicates that they do not specifically define the words "accident" or "crash" within their traffic statutes either.

The term "accident" has an implied, common definition that approximates the definition of motor vehicle accident contained in the *Uniform Vehicle Code* of 1926. The distinction in definitions becomes important to you, as a company safety representative, when you investigate certain types of highway crash events.

Consider a hypothetical situation in which your company driver makes a lane change, causing a nearby driver to respond by swerving their vehicle away from your company vehicle. The other driver travels off the highway during an evasive maneuver attempt, without ever contacting your company vehicle. The other driver is then injured or killed, and their vehicle sustains physical damage. Has your driver been involved in an accident?

Consider another hypothetical situation wherein your company driver makes a left-turn maneuver onto a roadway in front of an oncoming driver who is distracted for some reason. That distraction does not allow the approaching driver to view your company's vehicle until "the last second," causing the oncoming driver to take a rapid and dramatic evasive action. That driver loses control of his

4 Introduction and Definitions

vehicle and strikes another vehicle or object. Someone in the oncoming vehicle is injured and property has been damaged, but there was never any collision with your company vehicle. Has your driver been involved in an accident?

The answer to both of the hypothetical situations is probably, "Yes." Your driver-employee and your company vehicle may be included in the accident investigation and noted in the official traffic accident investigative report as a "non-contact" vehicle. By definition, a non-contact vehicle is a traffic unit that is involved in the circumstances of an accident event, but which does not necessarily collide with any vehicle or object.

Federal Definitions

Several definitions contained in Title 49 of the *Code of Federal Regulations* directly affect the safety employee or supervisor of a fleet operation, transportation organization, or motor carrier business that utilizes commercial motor vehicles or has operations that are regulated by the Federal Motor Carrier Safety Regulations. The regulations relating to traffic crashes have also been modified in recent years to allow more uniform record keeping and improved statistical reporting for federal governmental agencies, consistent with these definitions.

Federal Motor Carrier Safety Regulations are contained within Subchapter B, Chapter III, Subtitle B of Title 49—Transportation of the *Code of Federal Regulations*. Several specific sections, including the following definition of the word "accident," are identified and reviewed here.

Sec. 390.5 Definitions.

Unless specifically defined elsewhere, in this subchapter:

Accident means—

(1) Except as provided in paragraph (2) of this definition, an occurrence involving a commercial motor vehicle operating on a highway in interstate or intrastate commerce which results in:

    a. A fatality;

    b. Bodily injury to a person who, as a result of the injury, immediately receives medical treatment away from the scene of the accident; or

    c. One or more motor vehicles incurring disabling damage as a result of the accident, requiring the motor vehicle(s) to be transported away from the scene by a tow truck or other motor vehicle.

(2) The term accident does not include:

    a. An occurrence involving only boarding and alighting from a stationary motor vehicle; or

    b. An occurrence involving only the loading or unloading of cargo.

Several words or phrases used in the definition of an accident are also defined individually in Section 390.5. Those words include:

- commercial motor vehicle
- highway
- interstate commerce
- intrastate commerce
- fatality
- disabling damage

A review of some of the individual definitions is warranted, as they may be of significance to you.

Fatality Fatality is defined as "any injury which results in the death of a person at the time of the motor vehicle accident or within 30 days of the accident." This suggests that, within 30 days after the traffic accident, a representative of the transportation organization, fleet operator, or motor carrier company may need to contact the injured person(s), or their representative, to ascertain that person's status to ensure that the person has not died.

The definition also possibly infers that if death occurs to someone who was involved in an accident, the death needs to have been related to injuries sustained during the crash.

Disabling damage Disabling damage, as defined in the same section, discusses the physical alteration of the appearance, condition, or structure of the vehicle. It states:

Damage which precludes departure of a motor vehicle from the scene of the accident in its usual manner in daylight after simple repairs.

(1) Inclusions: Damage to motor vehicle that could have been driven, but would have been further damaged if so driven.

(2) Exclusions:

   a. Damage that can be remedied temporarily at the scene of the accident without special tools or parts;

   b. Tire disablement without other damage even if no spare tire is available;

   c. Headlamp or taillight damage;

   d. Damage to turn signals, horn, or windshield wipers which makes them inoperative.

This definition implies that if your driver can drive the truck away from the scene of the accident, by perhaps pulling a front bumper away from a tire, using an adjustable wrench to remove a fender, or using a screwdriver to remove a damaged headlamp, the damage may not be considered disabling.

Consider a hypothetical situation in which a truck is driven after an accident a short distance away from the accident scene into a parking lot. At that point, a decision was made that the company vehicle would need a tow truck to remove it from that location. It could be argued that the accident did not produce disabling damage, as the truck was actually driven from the scene.

You should be cautious when reviewing a police traffic accident report regarding an accident involving your company vehicle. The accident report may show the other vehicle as being towed from the scene. Keep in mind that police may tow a vehicle from the accident scene for a reason unrelated to the damage sustained by the vehicle during the accident. Examples of this situation include towing a vehicle because its driver is being physically detained for an active and outstanding arrest warrant unrelated to the accident, or towing a vehicle because it did not have proof of insurance documents available for examination by the police officer or state trooper at the time of the accident. Some police departments, sheriff's departments, and state police organizations will not allow a vehicle to remain on the highway, or on the shoulders of a highway, after an accident event.

Highway Section 390.5 describes "highway" as:

...any road, street, or way, whether on public or private property, open to public travel. "Open to public travel" means that the road section is available, except during scheduled periods, extreme weather or emergency conditions, passable by four-wheel standard passenger cars, and open to the general public for use without restrictive gates, prohibitive signs, or regulation other than restrictions based on size, weight, or class of registration. Toll plazas of public toll roads are not considered restrictive gates.

This federal definition is interesting because it specifically includes private property that allows an open access to the area by the general public. In most state statutes defining the word "highway," private property is not considered a highway. For example, in Colorado, Wyoming, and Montana, the statutes define the word "highway" as "the entire width between the boundary lines of every way publicly maintained..." Traffic statutes adopted by New York define the word highway as "a highway, road, street, alley, avenue, thoroughfare, bridge, or public driveway which is open to public travel and publicly maintained."

Each state in which your company vehicle operates should be checked for specifics on what is classified as a highway because private property may be included. Under federal regulations, however, an accident event involving a commercial motor vehicle or other regulated highway vehicle that occurs on private property and results in a death, bodily injury, or disabling damage as specifically noted within the regulations would be an accident that occurs on a highway. Such private property highway locations would include facilities such as a warehouse, distribution center, shopping mall, hospital, or hotel resort that a company vehicle might traverse, use to pick up passengers, or utilize to make deliveries.

Some law enforcement agencies may not respond to traffic accidents in these locations, however, because they occurred on private property, not on a public highway as their respective statutory definition classifies the word "highway." So, you will be unable to rely on the police to investigate such a traffic accident.

Figure 1–2 shows an example of a private property location. The accident clearly involves disabling damage and may have caused personal injury to the driver. The accident typically would be reportable because it occurred at a facility that was open to the public without restrictive gates.

Figure 1–2 Single vehicle tip-over accident that occurred on private property.
Photo courtesy of David Lohf.

There are specific requirements under the Federal Motor Carrier Safety Regulations *in addition to* state and local motor vehicle statutes, ordinances, and regulations dealing with accidents. These requirements generally dictate that a report is required for any event resulting in personal injury or death. Additionally, accidents resulting in only property damage may need to be reported to authorities. Each state or local jurisdiction will have specific minimum values of property damage that necessitate an accident event report.

The motor carrier and commercial vehicle operator, of course, is responsible for knowing and complying with these additional state requirements for reporting a traffic crash or accident. Information regarding additional requirements relating to traffic accidents can be found within state statutes, local ordinances, and federal regulations. You can locate these easily on the Internet in many instances or in private publications and resources dealing with the fleet operator, transportation organizations, and the motor carrier industry.

2

Planning and Preparation

It may sound bizarre, but each commercial motor carrier company, transportation organization, and fleet operator should plan ahead for its next traffic accident. The reason behind this unusual statement relates to the need for company personnel, especially safety and/or transportation operations personnel, to be prepared for what needs to be done, who needs to be contacted, and which responsibilities must be fulfilled by the company when an accident happens. The necessity to prepare and plan for the next time that a company vehicle becomes involved in a traffic mishap relates to the many investigative and research activities that should be initiated, the reports and documentation that should be collected and generated, as well as the potential civil liability that may attach to the company as a result of the crash.

One of the first areas that you should evaluate is the corporate structure of the company. If you work for a large fleet operator, transportation organization, or commercial motor carrier company, there may be corporate policies, procedures, and regulations already in place that identify and specify who needs to be involved with the accident investigation, what needs to be done, and what resources are available for your use. If this is the case, you should review and understand those policies, procedures, and regulations thoroughly. During the stress and anxiety of a serious traffic accident situation, you should be knowledgeable and familiar with corporate documents and policies and follow them as you complete your duties.

The opposite end of the corporate spectrum is the small fleet, organization, or commercial motor carrier company. In this situation, you may be employed by a company that has only a few employees, and you may be responsible for several areas of the company's operation. There may be only one person to handle dispatch, operations, and safety for the company. The company may be small enough that you have only general and simplistic guidelines about what to do, who to call, and what resources are available. In this situation, you might consult with several outside resources, network with other transportation company safety personnel, and/or devise the policies, plans, and procedures that will be needed when a traffic accident occurs.

The situation between the previous two examples is the position for many commercial motor carrier company safety personnel. The company has some operational policies in place, some personnel who have experience handling traffic accidents, and some established resources to assist in the event they are needed. The framework for investigating a traffic crash properly may be in place, but specifics on policies and procedures may be minimal. The company's response varies on a case-by-case basis, as there is no formal plan in place and no continuity to their response to a traffic crash event.

An examination and review of some of the procedures, policies, and resources that could potentially be necessary in the event of a traffic accident, and an explanation of *why* they may be of assistance to your investigation, may be helpful.

Corporate Policies and Procedures

There are no specific government requirements for a fleet operation, transportation organization, or commercial motor carrier company to have formal or written policies and procedures for accident investigation. The lack of formal policies, however, may result in some difficulties while following or attaining compliance with federal or state governmental requirements and regulations related to the operation of a vehicle fleet, transportation corporation, or commercial motor carrier business. Establishing formal policies might also help to improve the corporate safety structure and reduce uncertainty and confusion when a traffic mishap occurs.

Policies and procedures related to the investigation of a traffic accident are probably found within a safety-related area of the company operations manual. Company safety regulations may incorporate an accident prevention program as one of the tools to ensure the safety of all of employees and visitors to the corporate facilities. The safety programs of most commercial motor carrier companies, fleet operators, and transportation organizations follow those of general commercial and industrial corporations, usually addressing rules and regulations that should be followed by employees, as well as visitors and subcontractors for the company. These rules are often subdivided into two categories:

General Rules Those rules and regulations that would apply throughout the entire facility and for any type of operational activity of a company employee.

Specific Rules Those rules and regulations that are applicable to a given task, assignment, area of the plant facility, or to a given type of equipment, such as a commercial motor vehicle.

By necessity, the policies and procedures need to be in compliance with all federal, state, and local regulations for the company's type of business operation. Regulations relating to the Occupational Safety and Health Administration (OSHA), Workmen's Compensation, the Environmental Protection Agency (EPA), and the Federal Motor Carrier Safety Administration (FMCSA) are just a few of the types required by the government.

The corporate policy may state that all accidents, whether industrial, occupational, or motor vehicle related, must be reported to the company. The reporting typically occurs to the employee's supervisor or, in a larger corporation, to a company safety department. That policy, however, may not specifically state that the accident event must be "investigated" (as opposed to "reported"), or it may be ambiguous regarding the extent and magnitude of that investigation process.

For example, the company policy may state that, "All traffic accidents must be investigated by a member of the safety department." That rule does not state or imply if someone from the safety department must travel to the actual accident scene, or if they may rely on information contained within the traffic accident report filed by the law enforcement agency authorized to investigate the event. Further, the rule does not imply *when* a response to the accident scene by a member of the safety department should occur. The policy does not state what materials and documents need to be collected or which activities, such as interviewing witnesses, taking photographs, or creating a scaled diagram of the accident scene evidence, must be performed.

The problem may be compounded if the transportation organization, fleet operation, or commercial motor carrier company has a large regional or national territory, as opposed to a local operation. Whatever policies or set of regulations may be developed by a transportation organization, fleet operator, or commercial motor carrier company, they must be reasonable for the type of operation.

The question you should be asking is, "What are we going to lose if someone from our company doesn't go to the accident scene?" Unfortunately, as discussed in Chapter 3, the safety representative from a commercial motor carrier company or fleet operation should not rely solely on the law enforcement investigation. An attitude that, "The police will get everything I need," will lead to failure. Additional information that may be required by the company from the accident

12 Planning and Preparation

scene, from the vehicles involved, from the witnesses, from the drivers, and from the passengers is too important to leave to chance. The police may not have the same interest in or perspective toward the accident investigative results as your company.

Corporate "In-House" Resources

A medium-to-large-sized motor carrier corporation may have several operational areas, or in-house departments, that can provide some assistance to you when you're charged with the responsibility of reporting and/or investigating a traffic accident involving one of the company's vehicles. These departments may have relevant records, documents, and information relating to the employee and company motor vehicle and may have previously established business relationships with equipment suppliers, vendors, and information resources that can be utilized during your investigation.

Administration Department

This department may have records relating to the insurance policy, insurance carrier contacts, lease agreements (for situations when a commercial motor carrier company, fleet operation, or transportation organization utilizes owner-operators or other subcontractors to transport their freight or passengers), and equipment purchases.

The administration department may also provide logistical support in the form of secretaries or personnel to conduct research and/or follow-up activities.

Corporate administrators may provide supervisory controls and direction to various employees who may be involved in the crash investigation. Administrators may also provide guidance on the nature and types of reports and documentation that may be generated from the crash event.

Human Resources Department

This department maintains employee records relating to residence location and next-of-kin information. It may provide verification of employment status and answers to insurance coverage questions for an injured employee. It may also assist in filing incident and claim reports with workmen's compensation carriers, OSHA inspectors, other insurance companies, and/or medical care providers.

Maintenance Department

This department maintains information and records about the equipment involved in the accident. It may provide documentation of the date of the last preventative maintenance work on the equipment; the date of the last annual inspection completed on the bus, truck, or trailer; any previous complaints about the operation of the vehicle and what was done to correct the complaint; or any safety recall bulletins issued for the motor vehicle equipment involved in the crash.

The maintenance department may also assist in identifying reputable repair facilities or towing companies able to extricate and/or tow the vehicle in a safe and cost-efficient manner. Maintenance shop personnel may be able to identify parts to be ordered for replacement of damaged components, and they may be able to provide temporary, on-highway repairs and safety assessment inspections of the damaged vehicle.

Dispatch Department

The dispatch department may be able to identify dispatch assignments given to the driver-employee, the location and time of the last drop-off of freight or passengers, the type and identification of the freight or product that was on the truck unit at the time of the crash, the destination and ownership of the freight currently on the truck unit, and any special instructions or precautions relating to the current load. Dispatch may also be able to assist with identifying the previous hours-of-service for the employee-driver involved in the crash.

This department may also serve as a "clearinghouse" for contact to and from the driver, for information about the crash that flows into the company from outside sources, or for initial notification of hazardous materials spill command centers, such as the National Response Center and/or the Centers for Disease Control and Prevention.

Legal Department

The legal department may provide support in the form of legal advice regarding how the company should proceed in a given situation. It may provide a conduit for handling media requests and community inquiries relating to the accident, and may provide legal representation of the company driver in situations when criminal charges are filed against him or her.

The legal department may handle follow-up contacts with the next-of-kin of an employee killed during a crash event. This follow-up should occur after the coroner's office has verified notification of the next-of-kin, due to the emotional and traumatic nature associated with the initial notification process. This department may also assist in contacting potential claimants involved in the crash event if additional information is needed.

Legal department employees may assist in filing regulatory reports and dealing with requests from enforcement agencies. They may also provide guidance on the nature of reports and documentation that may be generated from the crash.

Driver

The driver is one of the most important sources of information about the facts and circumstances of the traffic accident. The driver may be able to take photographs at the scene of the crash shortly after the collision, possibly even before the arrival of law enforcement personnel. The driver may provide information on traffic volumes, local weather conditions, and pre-collision operational characteristics or peculiarities of the claimant vehicle(s). The driver may also be able to identify and contact potential witnesses to the circumstances and events of the accident, recall statements made by other participants in the crash and note observations of the post-crash condition of other drivers and passengers.

The driver should be able to provide the names of the investigating officers and their law enforcement agencies, and will be the initial liaison with emergency response personnel in the case of a potential hazardous materials incident involving the truck and/or its cargo load. Until other employees have arrived at the accident scene or are in a position to assist with the crash event, the driver will be the initial contact person for the company.

Safety Department

Other employees within a safety department may have information concerning contacts and resources that would be of assistance during a particular type of traffic accident. They may have established business relationships with outside vendors and contractors who can provide services, equipment, and materials. Other safety personnel may have background information regarding certain highway routes, customer facilities, and locations.

Outside Resources

Regardless of the size of the commercial motor carrier company, fleet operation, or transportation organization, many activities cannot be completed by an employee or are not cost-effective for the motor carrier or commercial motor carrier company to perform itself. Additionally, there are many situations in which the expertise and/or equipment necessary to perform an activity are not available within the company. In these situations, it makes sense to outsource those tasks and activities to a private company or service.

An important factor to consider is that you should initiate these contacts and develop these business relationships *before* they are needed. During the stress and confusion of post-accident activities, you should not attempt to initiate contacts, negotiate prices, identify services parameters, and dictate response time frames from outside service providers. Although there are many reputable and outstanding companies that provide services which may benefit your company after a crash event, you don't want to be in a one-sided business agreement because you need immediate assistance. Remember that after-hour contacts during the night, weekends, or holidays may be difficult without an established business relationship, especially when you are dealing with the situation "long distance."

Traffic Accident Reconstructionist

This resource person or company will be able to assist you by focusing on the physical evidence of the crash event, as well as the specifics of "how" and "why" the traffic crash occurred. The reconstruction company should have established experience in the investigation and reconstruction of highway traffic accidents in general and commercial vehicle crashes in particular.

Someone with practical "hands-on" experience, specific training, and industry-related knowledge (such as a law enforcement background) may be of greater value than someone from the engineering or academic sciences, who often lacks actual field experience in identifying, documenting, and collecting physical evidence and data at an accident scene. The reconstructionist may be accredited by the Accreditation Commission for Traffic Accident Reconstruction (www.ACTAR.org), an independent organization that develops and administers minimum-level competency examinations to persons engaged in the professional field of traffic accident reconstruction.

Insurance Claims Adjusters

This resource will be able to assist you in a variety of ways. The insurance adjuster may be able to provide outside referrals for traffic accident reconstructionists, clean-up, Hazardous Material (HAZMAT) remediation services, as well as local legal counsel. The company's insurance policy may provide financial support for some or all of these services, as well as out-of-pocket expenses incurred by your driver.

The claims adjuster can assist in interviewing drivers, intervening on behalf of your company with potential claimants or legal representatives of those potential claimants, and interviewing independent witnesses to the accident event. The claims adjuster can explain coverage issues within your policy and can assist with claims filed against your commercial motor carrier company by other insurance carriers.

Load Clean-up Services

This resource can help off-loading freight and product from your truck and/or trailer. A clean-up service may also provide storage for freight that can be salvaged from the accident scene or disposal of damaged freight at a later time.

Hazardous Materials Clean-up and Remediation Services

This resource will focus on containing, collecting, and neutralizing hazardous materials or fuel spillage from any of the vehicles involved in the crash or from freight or products the company vehicle was transporting at the time of the accident. The company will handle damaged freight or product from the truck unit, remove materials from the scene, and provide appropriate site clean-up and remediation of any contaminated soils, buildings, and property that resulted from the accident, as shown in Figure 2–1.

Figure 2–1   Hazardous materials clean-up service.

The HAZMAT remediation company may act as an intermediary with law enforcement personnel in charge of the accident scene and with environmental enforcement personnel. It may also assist your carrier with the compliance of federal regulations contained within Title 49, C.F.R. Parts 390 and 171 of the Hazardous Materials Regulations of the U.S. Department of Transportation.

Towing Services

This resource will provide extrication of your equipment from the wide range of post-accident, "at rest" positions in which your equipment may be found. Proper and safe recovery and removal of your truck or fleet equipment may mean the difference between the quick return of your company vehicle to the highway or a "total loss" of your equipment and its cargo. The towing company will remove your vehicle or truck from the accident location, provide secured storage of your equipment, and may assist with the transfer of your equipment to a repair facility or to your terminal facility (see Figure 2–2).

Figure 2–2 Post-accident recovery of a truck. *Photo courtesy of Colorado State Patrol.*

Maintenance and Body Repair Facilities

Local repair facilities may be able to provide services that would allow your driver to continue with his or her trip, once the necessary replacement of parts, repairs, and safety reinspection of your vehicle equipment have been completed. They may be able to repair a van, bus, truck or trailer component temporarily until the commercial vehicle returns to the home terminal facility.

Legal Representatives

This resource can be vitally important, especially if there is any potential liability exposure for the company or a potential for criminal charges that may be filed against your driver, and/or if personal injury or death occurred as a result of the collision. Most fleet operations, transportation organizations, and commercial motor carrier companies do not have the luxury of an in-house corporate attorney that is available 24 hours a day, seven days a week. They must rely on an outside legal counsel to provide legal assistance to the company.

An attorney can provide legal information, guidance, and advice on how to handle the variety of questions and situations that arise from a traffic crash. The attorney or legal firm should have experience representing fleet operation, transportation, and commercial motor vehicle accident cases, due to the many specific issues that often involve these types of cases. If the need arises, local counsel for your transportation company may be able to provide referrals for out-of-state attorneys.

Leased Equipment and Owner-operators

This resource may provide personnel or equipment support to your company if the fleet vehicle, bus, truck, or truck tractor and semi-trailer is owned or operated by an outside lease company or by an owner-operator contracted with your company.

Medical Facilities

Local medical facilities may provide post-accident drug and alcohol testing of your driver (as required by federal regulations in Title 49, C.F.R., Part 382), as well as follow-up medical services for him or her.

3

Tools, Equipment, and Supplies

The equipment and supplies you acquire allows you to complete tasks and conduct activities necessary during the investigation of a traffic accident involving a commercial motor vehicle. Even the simplest of tasks will require some type of "tools," even if those tools are only pen and paper. The purpose of this section is to identify the tools, equipment, and supplies you might need at the scene of a traffic accident or while conducting follow-up activities relating to a traffic accident. Once you have identified the items you may need, collect the tools and supplies before you need them.

Start your collection by reviewing the following lists of tools, equipment, and supplies. Obviously, depending on your situation, you may want additional items that are not included on these lists. Some items may not be necessary for certain accident situations, vehicle configurations, or areas of the country.

Photographic Equipment

- Still camera (digital and/or SLR film format)
- Video camera (digital or analog)
- Film supplies for SLR camera (various speeds of film for daylight to low-light situations: ASA 200, ASA 400 and ASA 800)
- Memory cards for digital camera
- Back-up camera (point-and-shoot, disposable, and/or basic SLR)
- Supplemental off-camera flash
- Assorted lenses, if not using an adjustable zoom lens
- Tripod
- Lens filters, especially a polarizer filter
- Closeable plastic bags to cover camera during inclement weather and to hold exposed film or memory cards
- Battery supplies or battery pack for camera and flash

Recording Devices

- Notebook, writing paper, or tablets
- Clipboard and/or notebook
- Envelopes or folders to hold papers, business cards, newspaper clippings, and documents
- Binding clips to hold papers on clipboard in windy conditions
- Hand-held tape recorder and tapes
- Writing instruments (pens, pencils, colored pencils)
- Fine-tipped permanent markers
- Assorted drawing guides and accident templates

Measuring Tools

- Measuring tapes (typically 25', 30', 100', and 200')
- Measuring wheels
- String line level and string line
- Plumb bob for identifying vertical location measurement
- Carpenter's level

Reference Resources

- Title 49 of the *Code of Federal Regulations* reference book
- Corporate contact telephone/cell phone lists
- Outside resource contact lists
- Palm- and computer-based address books
- Local and regional maps
- Cell phones and charger units
- Company reports and forms
- Accident check lists

Safety and Protective Devices

- Safety vests for all company staff who might be on scene (see Figure 3-1)
- Yellow flashing warning lights for vehicles stopped near the road
- Warning signs to advise approaching motorists of your presence
- Hazardous materials protective gear
- Traffic cones (12" or 18" with retro-reflective collars)
- Biohazard and blood-borne pathogen protection
- Latex gloves and work gloves
- Hard hat
- Safety glasses
- Small-quantity spill clean-up supplies
- Hand sanitizer and wipes

Miscellaneous Supplies

- Inclement weather gear, boots, and gloves
- Cold-weather clothing
- Magnifying lens
- Stopwatch for timing traffic
- Marking devices for highlighting highway evidence (fluorescent-colored spray paint, lumber crayons, sidewalk chalk)
- Closeable plastic bags for storing small evidence items from the vehicle and/or the accident scene
- Large plastic bags for holding larger items that may be soiled or contaminated by blood or fuel spills

22   *Tools, Equipment, and Supplies*

Figure 3–1   Safety and protective devices.

- Spray cleaners and rags for localized clean-up
- Assortment of mechanic tools
- Duct tape and masking tape
- Adhesive labels or identification tags for marking evidence and film
- Creeper or waterproof floor mat for laying on the ground
- Short step ladder
- Scaled model cars, trucks, and trailers

Scaled model cars, trucks, and trailers can be an asset when interviewing drivers, witnesses, and passengers (see Figure 3–2). They will also be helpful to you when trying to determine vehicle impact configurations, interpreting tire marks evidence, and conducting post-accident evaluations.

All of these items need to be stored in easily transportable containers or cases, such as hard-sided tool cases, hard-sided briefcases, storage tubs, and sports equipment bags. Cases and containers should be weather-resistant and durable.

If you have to respond to an accident location via a commercial airline, identification labels are necessary. The items should be properly packed to prevent damage to them while in transit and stored to allow easy access to those who might need them. The transportation of pressurized and flammable materials or products on a commercial aircraft is prohibited by federal regulations. Consult your commercial airline or the Federal Aviation Administration for additional

Figure 3–2   Scaled model cars, trucks, and trailers.

information on prohibited items. You may want to consider purchasing some of these products and supplies at your destination and then properly disposing of them after your use.

All your equipment should be ready so that you or someone else can "grab and go" with the equipment and supplies that are needed at the scene of an accident. The container(s) or case(s) should be stored in your company vehicle or at an office location easily accessible during normal business hours, nights, weekends, and holidays. If you work for a company with several people in your safety department, make certain that all safety employees know where the containers and cases are located. An alternative solution would be to create multiple equipment cases for several employees who may have to utilize the supplies for a traffic accident. These cases could be stored in company vehicles and/or the privately owned vehicles of safety employees.

A procedure should be established to replace or replenish any disposable or usable supplies or equipment. In some transportation or fleet operation companies, problems occur when tools and equipment are "borrowed" from the safety department supplies. For example, the camera, film, and flash may have been borrowed for use at a retirement party for a company secretary. After the party, the camera flash left on can cause the batteries to drain. In another example, all of the film may have been used to create a presentation for a training class. The camera may not have been put back into the case, but left in the dispatch office. Any of these problems can easily be avoided when you take a few minutes to replace, replenish, and check your supplies.

As you review the list, you can understand how these tools will allow you to collect and document information you gather and observe at the scene of a traffic accident. The equipment and supplies will allow you to be comfortable at the scene of a traffic accident during inclement weather and long exposure to extreme weather conditions. They will provide you and your co-workers some safety while you are working on or near a public highway and while you examine damaged vehicles. The tools and supplies will also allow you to conduct your accident investigation activities in a more efficient manner.

4

Law Enforcement Relations

If you respond to most accident scenes shortly after the occurrence of a traffic crash, you may have some contact with law enforcement personnel. The more severe the traffic accident, the higher the potential will be for encountering one or more officers from the state police, sheriff's department, and/or local police. The degree of friendliness, professionalism, and effective communication between law enforcement officials and the fleet operation, transportation organization, or commercial motor vehicle company officials during the investigation may determine how beneficial and accessible the resulting law enforcement information will be to you, to the company driver, and to the corporation.

To understand the nature of contact that occurs between the police agency and the transportation company safety or operations representative, a foundation needs to be established for what the "governing rules" will be.

Accident—Legal Definition

The genesis of most traffic laws in the United States is found within the *Uniform Traffic Code* of 1926. This document is the basis for many of the common traffic-related laws, ordinances, and legal enforcement authority, and it is utilized as the foundation for most state and local traffic regulations and statutes.

In almost all jurisdictions in the United States, a law enforcement agency is charged with the legal responsibility to investigate all traffic accidents. Webster's dictionary defines the word "investigate" as "to inquire with systematic attention to detail and relationship." This responsibility usually is derived from some statutory authority which dictates that a report of some sort will be created as a result of the investigation of all traffic accidents. Although this is a state-specific authority, the following is an example of those statutes.

> It is the duty of all law enforcement officers who receive notification of traffic accidents with their jurisdictions or who investigate such accidents at the time of or at the scene of the accident or thereafter by interviewing participants or witnesses to submit reports of all such accidents to the department on the form provided, including insurance information received from any driver, within five days of the time they receive such information or complete their investigation.
>
> —*Colorado Revised Statutes 42-4-1606(4)(1)(a)*

The traffic accident report is prepared to comply with a preestablished form developed by each respective state motor vehicle division that is responsible for collecting, maintaining, and collating traffic records and the resulting statistics generated from the traffic accident reports. Typically, the report is completed by one or more law enforcement officers assigned to "handle," or investigate, the traffic accident.

Law Enforcement Investigation

The role of the law enforcement officer (a local police officer, a sheriff's deputy, or a state trooper) at the scene of a traffic accident typically relates to several established activities. These activities include:

- Collecting information to complete a traffic accident report
- Rendering medical assistance to facilitating emergency medical services aiding any injured persons
- Removing any damaged property or vehicles from the accident scene
- Initiating any enforcement action relating to the traffic crash
- Clearing the accident from the highway so traffic can move through the area in a normal condition

How efficiently the law enforcement officer handles these basic tasks will relate directly to how much data, information, and documentation you will be able to collect. Many police departments, sheriff's departments, and state police agencies have highly trained officers and troopers who have substantial knowledge, training, and experience in the investigation of motor vehicle traffic accidents. There are some departments and agencies that have established teams of accident reconstruction technicians or major accident investigation teams that specialize in both catastrophic and commercial motor vehicle accident investigation and reconstruction and who investigate traffic collision events that involve fleet vehicles and commercial motor carrier vehicles. Typically, the results of these in-depth and detailed investigations supply a plethora of data and information regarding the traffic crash.

Some safety personnel in the transportation and motor carrier industry mistakenly assume that they, or someone from their company, do not have to respond to the scene of a traffic accident because "the police will get everything and I can just get what I need from the traffic accident report." Such a belief is often foolhardy.

A review of two accident report investigation results illustrates the potential inadequacies of some law enforcement investigations. The first report depicts a truck tractor and semi-trailer exiting from a truck stop immediately before it was "broadsided" (see Figure 4–1). The location of the area of impact was generalized and the diagram was prepared crudely.

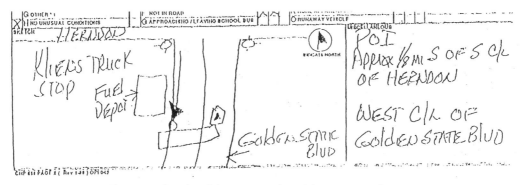

Figure 4–1 Crudely prepared accident report diagram.

The second report diagram depicts the investigative results of a 10-vehicle, rear-end crash on an interstate highway, which involved a truck tractor and semi-trailer vehicle (see Figure 4–2). Several people were injured severely and numerous vehicles sustained substantial physical damage from the crash. The officer did not take any measurements or photographs, did not indicate an initial area of impact, and incorrectly identified one of the highways. The officer was unable to provide any significant additional information regarding the accident when interviewed two weeks after the crash event.

Fortunately, these examples are exceptions, rather than the rule for the investigative results of police accident reports. In these days of diminishing financial resources for law enforcement agencies, however, the dilemma facing a law

28 *Law Enforcement Relations*

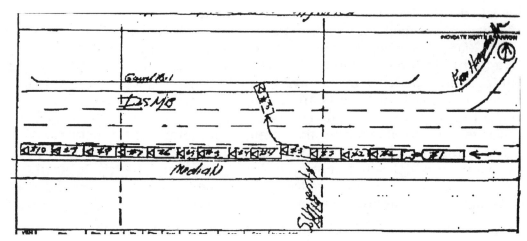

Figure 4–2 Incomplete accident report diagram.

enforcement agency attempting to maintain levels of service with a reduced level of staffing and financial resources is common. Some law enforcement agencies are reducing their time commitment to traffic accident investigation. This movement is mainly due to the quantity of time consumed during an investigation, the lack of manpower available to "handle" accidents, and the need to cover other calls for service. As a result, some city, county, and state law enforcement agencies have reduced their commitment to the thorough investigation of traffic accidents.

In fact, their commitment may be reduced to the point where the police agency may not respond to a traffic accident if it occurs on private property, if there are no obvious injuries that require emergency medical response, if there is no substantial property damage to any of the involved vehicles (usually ascertained by an estimate of the value of the damage to a vehicle), if there is no evidence of drugs or alcohol on the part of any driver, or if there is no need for a tow truck to remove a vehicle from the highway or street.

The result may be an incomplete investigation, a minimal investigation, or a "counter report." A counter report is filed by each driver involved in the accident at a local law enforcement agency. These reports are the only reports filed, as there is no independent investigation of the traffic accident by a law enforcement officer. Such reports fulfill the statutory obligations of both the respective driver and the law enforcement agency to submit a "report" regarding a traffic crash.

A lack of sufficient financial resources for law enforcement agencies may create another problem, affecting the training and competency of the investigating police officer, sheriff's deputy, or state trooper assigned to investigate the traffic accident or assist with the accident investigation. When budgets are tight in the public sector, one of the first areas to be reduced typically relates to the advanced training of law enforcement officers, especially in the nonmandated and specialty-focus areas, such as traffic accident investigation and reconstruction.

Figure 4–3 State troopers examining the highway surface for tire mark evidence.
Photo courtesy of Colorado State Patrol.

Figure 4–4 Local police officer collecting information for an accident report.

This may result in police officers, sheriff's deputies, and state troopers who may have minimal training opportunities that allow them to properly interpret physical evidence and understand vehicle dynamics relating to a collision event involving a fleet vehicle or a commercial motor vehicle. The officer may overlook the physical evidence present on the highway, misinterpret the physical evidence created during a traffic crash, or improperly assign fault for the accident based upon improper applications of scientific principles—all to the detriment of the transportation organization, fleet operator, or commercial motor carrier company. Fortunately, most law enforcement officers are trained adequately in traffic accident investigation and have the skills, knowledge, and experience to complete an investigation of most traffic accident events successfully.

The quality of the law enforcement investigation of the traffic collision has a direct and substantial effect on you—the fleet operator, transportation corporation official, or commercial motor carrier company safety person. The occurrence of a traffic accident involving a motor carrier's driver and vehicle has a high potential for significant financial exposure. The importance of a thorough and impartial accident investigation can be critical to the short-term and long-term financial exposure for the company, the safety record of the driver, and the safety rating for the company. Frequently, the fleet operator, transportation organization, or commercial motor carrier company relies heavily on the data entries in the traffic accident report for developing information regarding the "who, what, when, where, how, and why" relating to the traffic accident. The law enforcement agency's report may be the basis for an assessment of the liability exposure response the company will take in regard to any claims filed against it and the employment status of the professional driver. If the law enforcement agency has not been able to complete a thorough traffic accident investigation, there may be a significant lack in the quality and quantity of evidence and information available for analysis.

The issue is further compounded when a fleet operator or commercial motor carrier company faces a legal challenge in a civil law context. Although the laws may be the same, in many states the civil requirements for showing negligence are very different. Many states have civil laws that utilize a "preponderance of evidence" as opposed to the requirement of "beyond a reasonable doubt." Legal negligence can be established by the adoption of "joint and several liability," as well as "comparative negligence." Therefore, from the fleet operator's or commercial motor carrier company's perspective, the issues of concern to law enforcement are usually different and more limited than the issues that may evolve in the civil negligence arena.

Traffic Accident Report

Understanding the report completed during the law enforcement investigation is an important step in successfully completing your incident investigation. Your ability to understand how the information in a report was gathered and why it was included in a report, as well as what was *excluded* from a report and why, can be an important skill. Understanding the procedures the officer must follow when completing the report, understanding what information the officer can include in a report, and interpreting the coded data on the report will go a long way in assisting you to develop information regarding the traffic crash in question.

Each state motor vehicle department has developed an official traffic accident report form that must be completed for every traffic crash that has occurred within the state boundaries. The report format and information will be based, in part, on statutory definitions relating to the reporting of traffic accidents, the administrative policies of individual departments that will utilize the report, and regulatory guidelines created to assist the officer in completing the report. Each state will have a slightly different way of reducing the information developed through the investigation of a traffic crash into report format.

A review of reports from all 50 states, however, reveals some marked similarities, due in part to the requirements of federal agencies (such as the National Highway Traffic Safety Administration) involved in traffic safety and quasi-governmental advisory bodies (such as the American Association of Motor Vehicle Administrators). These advisory bodies and agencies develop the basic requirements for a traffic accident report, which is then tailored to the informational and statistical requirements of each state. Common areas on a traffic report relate to:

- Information on the date, time, highway location, city or county jurisdiction, and investigating agency
- Information on the identification and contact data of each driver
- Information regarding the identification of each vehicle involved in the accident
- Information relating to the road, weather, and traffic conditions present at the time of the accident
- Information regarding the movement of each vehicle before, during, and after the collision
- Summary of the results of the officer's investigation
- Information regarding enforcement action, if any, taken against a driver
- Information on the investigating officer and law enforcement agency

The following examples of traffic accident reports will illustrate the similarities and differences between state accident reporting formats.

Figure 4–5 shows the State of Colorado Traffic Accident Report. There is no place for witness information. The space for illustrating the physical damage for each vehicle is large and user friendly. Additional coding boxes are present on the right and left edges of the report.

Figure 4–6 shows the State of Utah Investigating Officer's Report of Traffic Accident. The format for driver and vehicle information is vertical, as opposed to side-by-side. The area used to identify physical damage sustained by a vehicle in a

32 Law Enforcement Relations

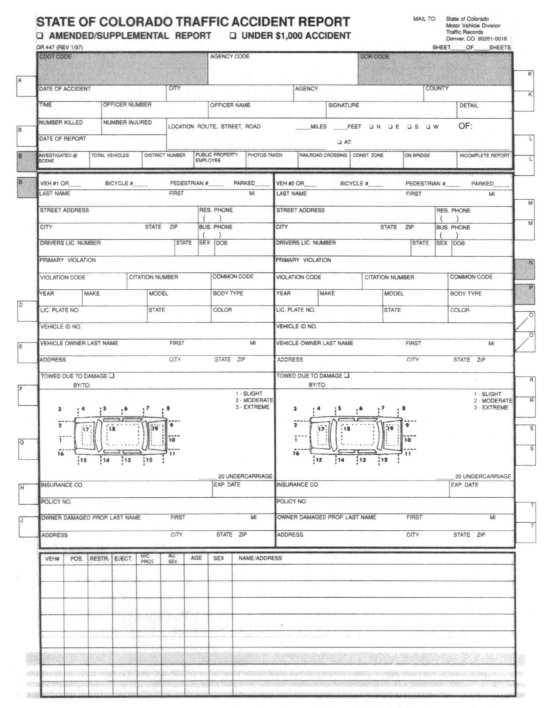

Figure 4–5   State of Colorado Traffic Accident Report.

crash is much smaller. The portion of the report used to identify the location of the traffic crash is more specific. Coding boxes are present on the right and left edges of the report.

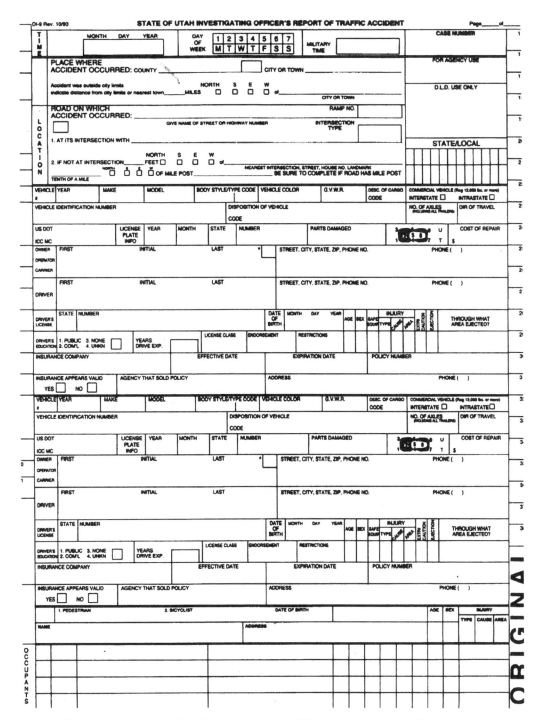

Figure 4–6 State of Utah Investigating Officer's Report of Traffic Accident.

The coding overlay sheet for the Police Accident Report from the New York State has numerous areas where coding information for conditions and facts regarding the accident can be recorded (see Figure 4–7). It is also important to ascertain what was *not* recorded by the investigating officer. For example, the opportunity for the officer to record "fatigued" for a given driver would be available. If that

34 Law Enforcement Relations

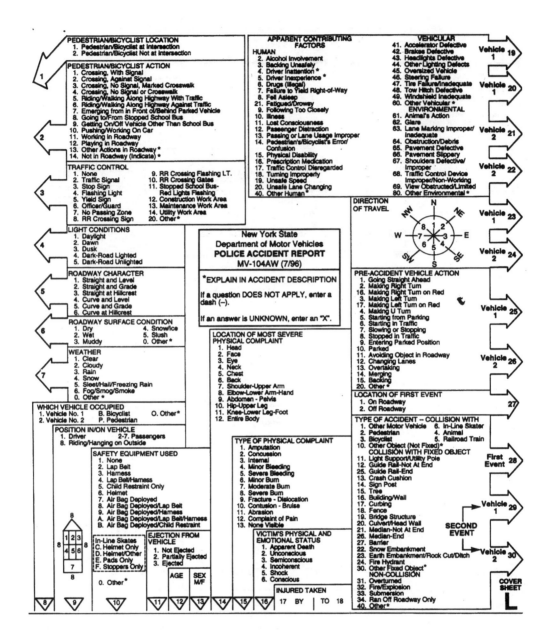

Figure 4-7 New York State Department of Motor Vehicles Police Accident Report coding overlay sheet.

coding selection was not made, then there would be an inference that the officer's investigation did not suggest that driver fatigue was an apparent contributing factor for the crash.

Working with Police at the Scene of an Accident

Many stressful activities occur at the scene of a major accident. The first priority of responding officials may be something other than assisting your company driver. Stabilizing the traffic accident scene, rendering medical aid to all of the injured parties, and ensuring that all persons involved in the collision are accounted for are the top priorities of responding officials.

Figure 4–8   Medical assistance for the injured is the first priority at an accident.
Photo courtesy of Colorado State Patrol.

Upon notification of a traffic accident involving one of your company vehicles, your immediate response should be to travel to the scene of the accident, if possible. This way, you can assist your driver (if necessary); provide protection of the load, the passengers, or the product contained in your company vehicle; ensure that towing and/or recovery of your vehicle does not further damage the equipment or other property; and obtain information about other drivers and vehicles involved in the crash. Your response to the collision site may also allow you to obtain additional documentation of the traffic accident by means of observation, video, photography, note taking, and measurements of the highway and other physical evidence at the scene.

One of the hurdles you will have to overcome, however, relates to establishing a working relationship with the law enforcement agencies and officers present at the scene and responsible for the investigation of the traffic collision. Remember that a traffic crash is actually a "crime scene." Police officers, deputies, and/or state troopers have the legal authority and obligation to maintain the integrity of the crime scene, to prevent contamination and spoliation of the evidence created by the collision, to collect or document the information in order to prepare reports, and to keep unauthorized persons away from the area.

At many accident scenes, the area surrounding the impact location can be chaotic. Law enforcement agencies that respond do not want to create more problems. They are committed to stabilizing the accident scene, creating a safe environment for other vehicles using the highway, maintaining a safe working area for responding emergency personnel, and providing control over what occurs at that location. So what can you do, as a representative of a fleet operator, transportation organization, or commercial motor carrier company?

Give the impression that you are a safety professional by acting and appearing as a professional. Wear a company uniform or jacket, if available, to identify yourself, along with safety vests, hard hats, and/or protective clothing.

- Act and communicate with law enforcement in a professional manner
- When you approach the accident scene, don't park your vehicle in an unsafe or hazardous location. Do not create more problems for the police clearing traffic from the area or prevent the movement of other emergency vehicles into or out of the accident location
- Approach the first officer you encounter and identify yourself as a safety representative of the fleet operator or commercial motor carrier company. Ask if you might be able to enter the location of the accident in order to provide information about your company vehicle, your cargo, or the product on the trailer
- Indicate that you are willing to assist the law enforcement agency with what they may need regarding your company's commercial motor vehicle. Indicate that you would like to check on your driver, if possible
- If you are allowed to proceed, ask that officer to identify the "officer in charge" of the accident scene, so that you can introduce and identify yourself to that person. Make contact with that officer, renewing your willingness to provide information and assistance if they need it
- Ask if you can answer any questions for them about your commercial motor vehicle equipment, your driver, or the load. This is especially important if there is a hazardous material product on your truck or semi-trailer that might require additional examination, monitoring, and/or product removal
- Ask if you can talk with your company driver
- Do not attempt to move the motor coach, truck, or truck tractor and semi-trailer combination unless you are specifically directed to or authorized by an officer
- Do not enter the cab or sleeper berth portion of the truck tractor or start removing items. The police may believe that you are removing contraband or tampering with equipment or documents inside the vehicle

- Ask if it is permissible to take photographs, with the understanding that you won't interfere with the police investigation or activities. Obtain the photographs or video of the accident scene without creating a hazard, interfering with the police officers, or compromising their investigation
- Follow the same unobtrusive procedure if you decide to obtain measurements. Ask the officer if it is permissible to take measurements or to complete a sketch of the accident scene
- If yellow crime scene tape stating "Crime Scene—Do Not Enter" or "Police Line—Do not Cross" is stretched across an area, do not enter that area unless you receive specific permission from one of the police officers

By using common sense and a professional demeanor, you should be able to remain in the accident area and accomplish what you need to while the law enforcement officers are completing their job. If you are asked to leave or you are not allowed into the accident area, do *not* disobey the police command. You quickly may find yourself arrested or ticketed for interfering with the police. If that happens, you will be of no value at the scene of the accident to your driver or to your company.

One of the benefits to responding to the accident scene while law enforcement officers are there relates to conducting your scene evaluation. In many situations, the highway is closed for an extended period of time to allow for the removal of damaged vehicles, to assist the injured, and to facilitate clean-up activities that often result from a commercial motor vehicle accident (see Figure 4–9).

Figure 4–9 Taking photographs of accident scene evidence is easier and safer when the highway is closed by the police. *Photo courtesy of Colorado State Patrol.*

Take advantage of the situation at the accident scene and take your photographs, measurements, and/or video. Obviously, you will never again have the opportunity to take detailed photographs and measurements of tire marks and collision physical evidence with the absence of moving traffic traveling through this area.

5

Investigative Factors

The process of investigating a motor vehicle crash involves identifying information and data about the accident with respect to three different subject areas: the human, the vehicle, and the environment. The purpose of this investigative process is to identify and document the myriad of contributing circumstances or factors present to create the crash event. Studies have shown that there are rarely singular causes for a traffic accident. They are caused by a combination of activities, factors, omissions, and mistakes. In order to investigate a traffic accident accurately, you need to understand the relationship between the various elements that join together and result in a crash.

In an attempt to articulate and categorize the components of a traffic crash incident, the National Highway Traffic Safety Administration developed an organizational chart of the three contributing elements. That organizational chart also identifies time intervals related to the accident event, which allows the investigator to identify and understand the complex issues and factors that combine to describe the manner in which the collision occurred and the reasons for the crash.

The three elements include the human aspect, the vehicle equipment, and the environment, which refers to both the highway environment as well as the atmospheric environment. Combined, they form a relationship often referred to as "H-V-E." These H-V-E elements and the three time periods for studying those elements, provide a platform for assembling information regarding a traffic crash, which allows you to develop conclusions about the cause of the accident. Compare the three elements of a traffic accident with the three legs of a stool. Each leg provides support and stability to prevent a situation or an event from occurring. When one leg of the stool does not perform as expected, something happens. We call it an "accident."

The Human Element

- The human involvement in an accident includes people such as the drivers involved in the event and the passengers who may occupy any of the vehicles involved in the event.
- The human involvement also includes any independent witnesses to the crash event or its aftermath. Witnesses may be passengers (such as those on a city transit bus), other motorists in the immediate area of the crash, motorists who viewed the driving behaviors of one of the involved motorists before the crash event, or bystanders who may have been walking or standing near the crash location.
- The human element also includes people who have some relationship to the accident event, such as police officers, fire and emergency response personnel, ambulance and paramedic responders, tow truck drivers, HAZMAT clean-up responders, independent claims adjusters, independent photographers, and news media personnel who responded to the accident scene. Any of these people may be able to offer information on the activities, events, and circumstances surrounding the accident.

The Vehicle Element

- The vehicle element includes each vehicle involved in the traffic accident. The year, make, model, seating capacity, engine size, transmission type, and ownership information are initial points of data that need to be collected.
- You also need to focus on the dimensions, the design profile, and the presence of cargo in each vehicle. These areas will allow you to evaluate, to some extent, how the vehicle responded during the pre-impact approach movement and how the vehicle reacted during the crash event.
- The vehicle element also relates to the presence of, or the absence of, physical damage on each vehicle. This information will allow you to assess the manner in which the vehicles collided, the magnitude of collision forces, and the potential for personal injury to occupants within each vehicle.

The Environment Element

- The environment element includes the highway environment as well as the ambient and atmospheric conditions present at the time of the crash.
- The highway environment includes the type of pavement surface, the condition of the pavement, contaminants or debris on the pavement surface, the various types of traffic controls in the area, and the presence of artificial light sources. Terrain, topographical features, and land use in the area where the highway is situated are also included in this element.
- Weather conditions, such as wind, rain, snow, sleet, and fog, are factors incorporated in the environment element. Lighting conditions related to the extent of daylight or the degree of darkness and the influence of fog, wind, rain, and snow on the driver's ability to see are environmental elements.

Time Frames

During your investigation and evaluation of the traffic accident, you must document and organize data, information, and evidence relating to these subject areas during three different time frames:

- Before the crash event
- During the crash event
- After the crash event

Keep in mind that the time frames are not necessarily narrowly defined. For example, "before the accident" could encompass several minutes, several hours, several days, or several months prior to the collision event. Determination of the extent of the time frame is entirely up to you, the nature of the accident event, and the depth of your investigation.

Identifying contributing circumstances and causal factors during three different time periods can be simplified by using a very basic spreadsheet. Your task should involve identifying and organizing information and data within each of the spreadsheet cells. In some crash events, however, your focus may be widened with the addition of certain cells. Table 5–1 illustrates how the elements for investigation integrate with the three different time frames.

In Table 5–1, only four items are listed for each accident component category at each different time interval. Your investigation should not be limited to just those topics and subtopics. These examples are listed to assist you in understanding the concepts and to recognize the unlimited potential for inquiry, if the circumstances of the traffic accident necessitate additional or in-depth inquiry.

Table 5-1: Time Frames and H-V-E Elements

	BEFORE CRASH EVENT	DURING CRASH EVENT	AFTER CRASH EVENT
HUMAN	Physical Condition Medical Limitation Driver Inexperience Fatigue	Intoxication Distraction Unfamiliarity with Area No restraint	Personal Injuries Impact with Interior Medical Treatment Statements Made
VEHICLE	Vehicle Specs Maintenance History Unsafe Loading Modifications	Angle of Impact Approach Speed Lane Position Load Shift	Exterior Damage Interior Damage Occupant Ejection Component Failure
ENVIRONMENT (Highway and Atmospheric)	Highway Design Traffic Control Device Surface Treatments Traffic Volume	Weather Conditions Travel Advisories Tire Marks Gouges	HAZMAT Spill Towing and Recovery Highway Fixture Repair Load Clean-up

Let's examine a hypothetical accident situation to illustrate how you can expand your investigation of the category "Human—Before Crash Event." Our example involves a fairly new employee to your company who was involved in a traffic accident in the morning near a customer location. Your inquiry into the investigative category may include the following questions:

- How long has this driver been with the company?
- How many times has this driver traveled this route or delivered to this customer?
- How much experience did this driver have with the type of company vehicle, truck, motor coach, or truck tractor and semi-trailer that he was driving at the time of the accident?
- What training did this driver have when he became an employee?
- Who performed the road test of this driver before he or she was hired?
- Did the road test cover driving situations similar to those that occurred during the accident event?
- What type of driving experience (OTR, local, regional, type of truck equipment) did this new employee have prior to being hired?
- What type of experience did this driver have hauling your type of products and trailer loads?
- Did this driver suggest or indicate a medical situation related to the accident event which was not disclosed on his application or during the physical?
- Did this driver have anything in the cab, such as food or a cell phone, that may have caused a distraction?

- Was this driver given proper directions, or did the driver inquire about possible routes to the customer location?
- Does this driver have any hobbies, family situations, or outside activities that may have caused unusual fatigue?
- Did any of the driver's coworkers notice any problems or unusual behavior that may suggest a situation or influence outside the company that significantly affected his or her ability to work safely?

As you can see, it would be fairly easy to expand the focus of your investigation into the category of "Human—Before Crash Event," if the circumstances or accident situation warrant it. With the examples cited, it is clear that you should be considerate of an employee's interests and activities outside of the work environment. Depending on the circumstances, however, you may have an indication to investigate further.

Elements of a Traffic Crash

Typically, common elements can be identified in traffic collisions. These elements occur prior to, during, and subsequent to the crash event. In many situations, the exact location or area where some of the events transpire may not be readily identifiable. A discussion of the common elements may be helpful in understanding the entire sequence of events that can occur in a crash.

Possible Perception

This element refers to the area or general location where a hazard, or potentially hazardous situation, could be perceived by an attentive person. Possible perception typically occurs prior to perception. The driver's ability to view the area ahead allows an opportunity to understand the situation, evaluate options, and then decide on a course of action. This process of possible perception may take place over an extended time or it may occur with perception.

Driver training courses and guidelines stress the necessity for any driver to scan ahead of the vehicle's position. That distance is recommended to be approximately 5–15 seconds, with respect to the speed at which the vehicle is traveling. If a time duration of 10 seconds is assumed with a speed of 65 miles per hour (which can be calculated as an equivalent 95.2 feet of travel distance per second of time), this would require scanning a distance of approximately 950 feet, or slightly less than 0.2 mile.

Many state-issued driving manuals also recommend the practice of advance scanning, which is incorporated into defensive driving techniques. Your investigation should, if possible, document the appropriate range of distances for possible perception, so you can understand what a driver may have viewed during this phase preceding the accident event.

44 Investigative Factors

There are limitations on a driver's ability to scan ahead and to extend the time of possible perception. Factors relating to topography, highway geometry, ambient light, and traffic congestion may alter or reduce the ability to extend the possible perception element (see Figure 5–1).

Figure 5–1   Factors limiting the time of possible perception include topography, highway geometry, ambient light, and traffic congestion.

Perception, Reaction, and Response

These elements are combined, as there is no practical means to separate them. Some published materials suggest that a time duration of three-fourths of a second for the perception process and three-fourths of a second for the reaction-response process should be applied to pre-collision activities for drivers exposed to an impending traffic collision. These figures, which are used in many traffic books, safe-driving pamphlets, and state-issued driving manuals, have been used by many safety personnel when conducting their investigations.

However, current analysis and research conducted by Jeffrey W. Muttart of the Accident Dynamics Research Center identifies the substantial limitations imposed by such assumptions regarding perception, reaction, and response. In his extensive research and real-world testing, Mr. Muttart identified several factors that contribute to his conclusion that drivers will respond differently to different stimuli while driving and when faced with a potentially hazardous situation. The situational factors that may need to be considered and/or evaluated include the following:

Contrast Contrast of the potential hazard or situation related to the surrounding area. This is, in part, a function of the quantity of light available to view the potential hazard or situation.

Anticipation Anticipation of the driver, by analysis of whether the driver knows the potential stimulus and potential responses to that stimulus while driving.

Strength of the stimulus Strength of the stimulus, which relates to the movement, size, and intensity of the stimulus, indicating a potential hazard or situation.

Eccentricity Eccentricity which refers to the angle at which the driver is looking, relative to the location of the potential hazard or situation. This relates to both the separation of the potential hazard stimulus to the sides of the driver and the separation distance ahead of the driver.

Cognition Cognition which refers to the decision-making process by a driver based upon the information available to him or her, the complexity of that information, and the driver's recognition of the options available.

Response complexity Response complexity which refers to the ability of a driver to consider any alternative or optional responses to a given hazard or situation.

Mr. Muttart developed a computer-based analytical process for evaluating perception-reaction-response factors, wherein the factors and contributing circumstances for a given situation are itemized and then evaluated. He concludes that each situation must be analyzed before a range of probable perception-reaction-response times can be offered. His research, as well as the research of other analysts involved in the human aspects of traffic crashes, strongly discourages assigning an arbitrary assumption of a perception-reaction time to this grouping of accident elements.

Encroachment

Encroachment occurs when another vehicle or object enters the path, or intended path, of travel of a vehicle. Once that occurs, the offending vehicle or object quickly changes from a possible hazard to an imminent hazard, depending upon the time frame, the closure speed, and the separation distance between the vehicles.

46 *Investigative Factors*

The encroachment does not necessarily have to occur when a vehicle, for example, crosses the center line of a highway. It can occur at an intersection, a private driveway exit, or a highway interchange. Encroachment can also occur during a lane change situation, with both vehicles traveling in the same direction on a highway, or when one overtaking vehicle comes in close proximity to the vehicle being passed.

The proximity of the potential hazard may be very close, as in an undivided highway where two opposing directions of travel are present. Typically, vehicular traffic on these highways travels at speeds in excess of 40 miles per hour. Therefore, the time duration to detect and evaluate the encroachment may be minimal. The time duration may also be minimal, due to additional highway environmental factors, such as a grade and curvature in the roadway alignment. Figure 5–2 depicts the evidence from a head-on collision between a delivery truck that crossed over the center line and a light-duty pick-up truck.

Figure 5–2 Evidence from a head-on collision between a delivery truck, which crossed over the center line, and a light duty pick-up truck.

Start of Evasive Action

The start of evasive action refers to the location where physical evidence exists indicating where an attempted accident avoidance maneuver began or the location where calculations can be completed to indicate where that maneuver probably began. In most accident avoidance situations, a driver typically has limited options for avoiding the potential hazard or impending collision. Those options include:

- Steering toward the right
- Steering toward the left
- Braking the vehicle to slow or stop
- Accelerating the vehicle
- Implementing both braking and steering maneuvers
- Doing nothing

Keep in mind that significant evasive maneuvers do not always create physical evidence on the pavement surface. Thus, the absence of *evidence* of an evasive maneuver attempt does not necessarily establish evidence of the absence of an evasive maneuver attempt.

For example, during the approach to a traffic signal light that is changing from green to yellow to red, a driver may aggressively slow the vehicle by braking, but not leave any tire marks. A substantial braking maneuver by a driver operating a vehicle with an Anti-lock Braking System (ABS) may not leave distinguishable tire marks on the pavement surface. A swerve to avoid a hazard, such as an animal in or near the traffic lane, may not create tire marks to indicate where the swerve was initiated. For those types of evasive maneuver incidents, repetitive mathematical calculations may need to be performed, utilizing reasonable value ranges for perception, reaction, deceleration, acceleration, and/or swerving.

The accident shown in Figure 5–3 related to the encroachment of a vehicle from the intersecting highway on the right side of the photograph. A combination truck unit traveling on the through highway initiated evasive action by aggressively braking. A jack-knife eventually occurred with the combination truck unit. If physical evidence to indicate braking is present, as illustrated in the photograph, you can determine where that evasive action started relative to the collision location.

The arrow in the photograph indicates where tire marks begin on the pavement surface. It is important to note, however, that the actual braking process was initiated *prior* to this location. Considering the time and distance consumed by the commercial vehicle from the initiation of pressure on the brake, the time required to build system pressure to effectively slow the vehicle, and then the transition from rolling tires to fully sliding tires on the vehicle, the commercial vehicle initiated evasive action prior to the start of tire marks shown in the photograph. Studies have shown that this time duration could be within a range of approximately 0.25–1.5 seconds, depending on the mechanical set-up and condition of the brake system, vehicle design, loading considerations, and pavement surface conditions.

Figure 5–3 Evidence of evasive action. *Photo courtesy of Lafayette, Colorado Police Dept.*

First Harmful Event

By definition, the first harmful event relates to the first occurrence of personal injury or property damage involving the movement of a motor vehicle; this characterizes the collision type. The first harmful event is usually classified as one of the following traffic accident events:

- Non-collision on or off the roadway (such as a rollover crash)
- Collision with pedestrian
- Collision with other vehicle in motion
    - Broadside
    - Full impact—opposite direction (head-on)
    - Full impact—same direction (rear end)
    - Partial impact—same direction (sideswipe)
    - Partial impact—opposite direction (sideswipe)
    - Approach turn collision
    - Overtaking turn collision
- Collision with other vehicle
    - Parked vehicle
    - Bicycle, motorized bicycle, or skate board

- ○ Railway vehicle
- ○ Highway maintenance vehicle
- Collision with animal
- Collision with other object (highway fixtures and controls)

Initial Contact and Maximum Engagement

The time interval of *initial contact* relates to the point when a moving vehicle comes in contact with another vehicle or object. The time interval relating to *maximum engagement* designates when the two objects, having collided, attain the greatest penetration in damage, and when any momentum (mass multiplied by velocity) exchange between the two vehicles has been completed.

In Figure 5–4, the photograph on the top shows initial contact has just occurred. Maximum engagement has occurred in the photograph on the bottom, in a time duration of approximately 0.10–0.20 seconds. Note that the air bag has been deployed inside the Ford vehicle and the aggressive braking by the driver has produced significant forward weight shift within the vehicle, compressing the front suspension.

In Figure 5–5, initial contact is occurring in the upper left photograph. Maximum engagement is occurring in the upper right photo and initial separation has occurred by the lower left photo. Notice the movement of the semi-trailer, as evidenced by the white stripe painted on its tire sidewall. During the collision phase, the changing positions of both the semi-trailer and the Ford Explorer relative to the stationary camera position placed adjacent to the two colliding vehicles also indicate the forward movement, or translation, of the vehicles toward the right in the photographs.

Disengagement or Separation

This element occurs when contact between the two vehicles, or the vehicle and another object, ceases. The energy transferred from one vehicle to another, due to differences in the speed and/or weight between the two vehicles, has occurred. The post-impact velocity of each vehicle causes the *separation*, typically due to the residual speed of one vehicle after the actual collision phase (defined as the initial contact and then maximum engagement) has finished.

Figure 5–6 shows two vehicles involved in a staged collision that have separated. The initial contact and maximum engagement have already occurred. The damage resulting from the collision is evident on the front structures of both the white van and the pick-up truck. The white van is still experiencing the transferred kinetic energy resulting from the collision, as evidenced by its clockwise rotation and its airborne displacement (or movement from where the initial contact position was located). The time interval between initial contact and separation in this type of head-on collision is typically 0.10–0.30 seconds.

50 Investigative Factors

Figure 5–4 Initial contact (top) and maximum engagement (bottom).

Elements of a Traffic Crash 51

Figure 5–5 Crash sequence time intervals. Initial contact (upper left), maximum engagement (upper right), initial separation (lower left), final rest position (lower right).

Figure 5–6 Separation after initial engagement.

Final Rest Position

The point where a vehicle finally comes to a stop after crashing is referred to as the *final rest position*. This location is of importance, as it is usually the position noted, documented, and measured by law enforcement officers assigned to investigate the traffic accident. Typically, this will be the starting point of your investigation of the traffic accident.

The final rest position is also important as it may allow an assessment of the speed of one or more of the two vehicles at impact. The manner in which a vehicle arrives at the location of final rest can be categorized as either of the following:

Uncontrolled Movement to final rest was dictated by the forces of the collision without any input by the driver (see Figure 5–7).

Controlled Movement to final rest was directed and controlled by the driver of the vehicle (see Figure 5–8).

The significance of final rest positions relates to the information that may be derived from that data. The location of final rest and the manner in which the vehicle(s) arrived there may allow a determination of:

- Impact location
- Departure angle
- Approach path of vehicle(s)
- Inference of speed prior to impact
- Velocity change during and after impact

The location of final rest positions is determined by tire mark-evidence, metal scratches/gouges in the pavement surface, liquid or solid debris that has leaked or fallen from the stopped vehicle, a fluid trail from the vehicle as it traveled to that location, or markings made by the investigating officer on the pavement surface. Information may also be derived from interviews of drivers, witnesses to the accident, and bystanders who came upon the accident after it has occurred.

Pavement markings outlining and highlighting tire-mark evidence created by one of the vehicles in a multivehicle, head-on collision, can be seen in Figure 5–8. The physical evidence strongly indicates that the driver of this vehicle steered the vehicle off of the highway and onto the shoulder after the crash event. This driving control occurred even though a tire of the vehicle was obviously damaged. The arrow in the image on the right points to the "T" marking spray painted by the officer at the point where the left rear wheel of the vehicle came to a controlled stop.

Elements of a Traffic Crash 53

Figure 5–7   Uncontrolled positions of rest.

Figure 5–8   Evidence of controlled stop.

6

Recording Techniques

Two of the easiest ways to record information are to write down observations of what you see, hear, and learn about a traffic accident, and to record your observations and interviews by using a tape recorder. Minimal financial investment is required to obtain the tools for these methods of recording information. The data you record can be easily retrieved and transcribed, if necessary. Minimal skill is needed to learn to use the tape recorder, and it's typically small enough to carry around in your pocket.

A disadvantage to using a tape recorder as your sole source of documentation during a traffic accident investigation is that it can be time consuming to transcribe. It might also be difficult to retrieve a specific comment or note regarding a measurement without listening to the entire tape. However, the advantage of tape recording is that you can quickly record a substantial amount of information just by talking. Your thoughts can be recorded efficiently while examining a damaged vehicle, inspecting tire marks on a roadway, driving from the accident scene to the tow yard, or while talking to the investigating police officer.

The disadvantage to using paper and pencil as your sole source of documentation is that you cannot possibly write down everything you observe, hear, or measure during your accident investigation. The adage of "a picture is worth a thousand words" suggests that the time necessary to record on paper everything you may observe about the highway layout and appearance, and about the presence of physical evidence, will be extensive.

An alternative to solely using handwritten notes or tape recordings is to use that method as one of several methods to record information regarding the traffic accident. Just as your toolbox does not hold just one tool, you should have the tools for several methods to record information. You might think this is redundant, but it's a good idea not to rely on just one method to record data, as no single method works best. Your approach to documenting information should include paper and pencils, photographs and/or video, and measuring techniques to ensure that all of the information you gather can be utilized later, if needed.

It is important to keep in mind that *everything* you gather, create, develop, record, and write down can be discoverable in a legal proceeding. So whatever method or methods you utilize for documentation, you need to complete your investigation in a professional and careful manner. If you record, for example, your thoughts about what a driver may have been doing just before the crash, you should not use verbiage to suggest or indicate that, "This is what happened." Be cautious how you record the information and be aware of having the recorder activated when the conversations of others nearby could be recorded accidentally.

If you write something down, record "only the facts" that you have determined and not your unsupported opinions or guesses. Write legibly, so at a later time you can figure out what you wrote. Note the information in a systematic manner; do not scatter it all over the paper. Consider using pencils so that you don't have to scratch out misspellings and honest recording errors. Such an approach may save you from being embarrassed later.

Field Sketching

Another easy method of recording information is to prepare a sketch of what you observe at the scene of a traffic accident. This graphical depiction is often called a field sketch because it is prepared in the field, as opposed to an office, and because it is a sketch, as opposed to a scaled diagram. The field sketch supplements any photographs and/or video recording of the accident scene. The sketch or drawing that you prepare can be used to corroborate the photographic and observational information that you develop, as well as support portions of your analysis and conclusions regarding your investigation.

The field sketch can be as individualized and detailed as you require. The only limitation is your skill as an artist and draftsman. There are no strict requirements for a field sketch, as it is a tool that *you* prepare for your subsequent use. The field sketch allows you to record the overall general view of the accident scene as it appeared during the initial phase of the investigation. The field sketch may document the position of debris, vehicles, physical evidence from the crash, and witnesses. It also may depict information concerning an accident scene, a relevant traffic control sign or signal, the position of light sources, and/or the appearance of damage on a vehicle that you believe to be important to your investigation.

The information that is recorded in a field sketch can be as variable as the situation necessitates. You could sketch out, for example, the layout of a neighborhood or area surrounding the accident location, the appearance and placement of pre-impact tire marks leading to the area of impact, or the general layout of a curve or intersection where the accident occurred. There are no specific rules for what needs to be contained within the field sketch. Your might even consider preparing several sketches to illustrate different aspects of the collision event or a sequence of sketches of the physical evidence on a highway surface that extend over a large distance.

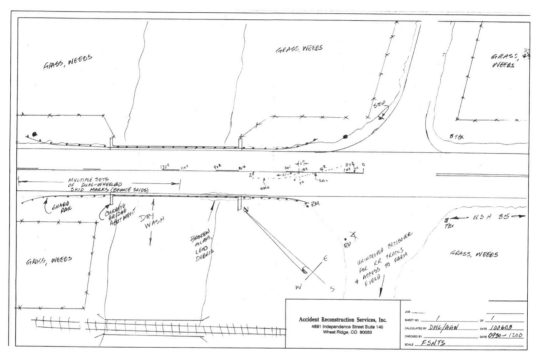

Figure 6–1 A quality field sketch.
Sketch courtesy of David Lohf.

The field sketch in Figure 6–1 illustrates the simplicity of a sketch, but also indicates the quantity of detailed information that can be included in the sketch. Some of the sketch was prepared with the assistance of a straight-edge or a ruler, and some of it was drawn "free hand." Physical evidence noted at the scene of the rear-end accident was incorporated into the highway geometry and layout.

The field sketch in Figure 6–2 illustrates what information you might consider incorporating from a traffic accident that occurred in an urban area.

58 Recording Techniques

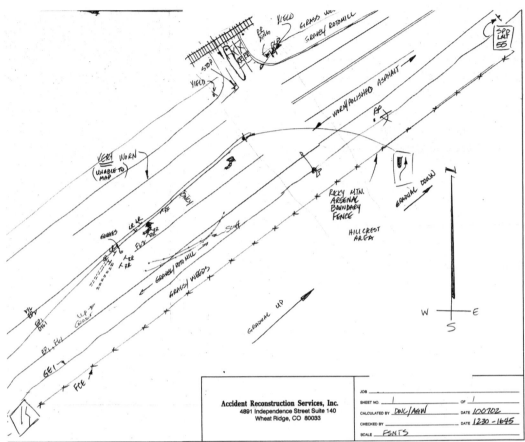

Figure 6–2 Field sketch with information about an accident.
Sketch courtesy of David Lohf.

Materials Needed to Prepare a Field Sketch

- Grid paper or plain paper
- Template or ruler
- Pencils and Eraser
- Colored pencils
  - Allow easier identification of different tire marks' lines
  - Allow quick recognition of different pavement lane lines
- Rigid writing tablet or clipboard

Items That Might be Included on the Sketch

- Identity or reference for the traffic accident
- Precise location of accident scene or location that is being sketched
  - City, town, or community; county; state
  - Route, highway, or street identification

- Street address of nearby structure, business, or residence
- Distance and direction to the nearest intersection or highway junction
- Distance and direction to a local permanent reference (i.e. bridge name) or mile post
- Highway/street layout and appearance
  - Pavement
    - Type of material
      - Bituminous asphalt, blacktop
      - Concrete
      - Gravel, dirt
    - Surface condition
      - New
      - Traveled
      - Traveled and polished
      - Rutted
      - Smooth
    - Surface irregularities
      - Patches
      - Potholes
      - Rough areas
    - Number of through lanes
  - Width of traffic lane(s)
  - Surface marking, type, and condition
  - Specialty areas
    - Left-turn and/or right-turn lanes
    - Median(s)
    - Painted islands
  - Shoulder areas
    - Surface composition
      - Asphalt
      - Gravel
      - Dirt
      - Grass
    - Surface condition at time of accident
    - Ditches adjacent to the roadway
    - Drop-offs or embankment cuts
    - Guardrails, curbs, and retaining walls
      - Type
      - Height

- Condition
- Location
- Obstructions near shoulder
  - Type
  - Location
  - Dimensions
- Traffic control devices
  - Signs
    - Type (identify all applicable)
      - Regulatory (stop, yield, speed limit, etc.)
      - Warning/Cautionary (curve ahead, stop ahead, etc.)
      - Informational (town and street names)
      - Directional (direction to a town or destination)
      - Guidance (points of interest, services)
    - Distance to area of impact
    - Visibility and legibility of sign
  - Traffic control signals
    - Type
    - Distance to area of impact
    - Operational sequence, timing
    - Visibility
  - Posted speed limits (maximum permissible)
    - Auto
    - Truck
  - Other (specify)
- Road design
  - Grades
    - Up
    - Down
  - Curves
    - Length
    - Degree (sharp, moderate, gradual)
    - Direction (right or left)
  - Straight segments (length of segment preceding area of impact)
  - Super-elevation (slope or curvature in side-to-side direction)
  - Sight distance limitations
  - Terrain of adjacent areas and development

- View obstructions and interferences
- Artificial lighting
    - Operational?
    - Location
- Glare source
    - From reflections of polished glass of building exteriors
    - From sun position at the time of the accident
- Drainage problems or inlets
- Construction
    - Activity
    - Work zone traffic control devices
    - Detours
- Unusual features
    - Bridges
    - Underpasses
    - Overpasses
    - Vegetation
    - Fallen rocks
    - Snow fencing

○ Environmental conditions occurring at the time of the accident
- Excessive temperatures
- Precipitation
- Significant wind velocity
    - Estimate on wind speed (gusting vs. steady)
    - Prevailing wind direction
- Natural light (light angle, if a possible factor: sunset or sunrise)

○ Accident evidence on highway infrastructure
- Area-of-impact evidence
    - Gouges
    - Scuffs
    - Skid mark change in direction
- Skid marks
- Tire-mark evidence created during the accident event:
    - Tire marks
    - Tire prints
    - Ruts
    - Furrows
    - Erasure

- - - Gouges
    - Scratches
  - Approximate position with respect to lane/road markings, curves, fixtures, etc.
  - Temporary evidence from the accident
    - Debris
      - Undercarriage dirt displaced from the vehicle(s)
      - Splashes, spills
      - Dribbles
      - Vehicle components
      - Load
- Vehicles
  - Path of travel approaching area of impact
  - Path of travel after impact
    - Uncontrolled
    - Driver-controlled
  - Point of rest location
- Other vehicles involved
  - Assign each vehicle a number
  - Identify their approach path
  - Identify additional impact locations, if possible
  - Identify points of rest
- Traffic
  - Volume
  - Direction
  - Evidence of evasive action by other vehicles due to crash
- Any other pertinent information

Coordinate Measuring System

The coordinate measuring system utilizes a format consistent with an X- and Y-coordinate system, with the Y-axis having a north-south alignment and the X-axis having an east-west alignment. All points to be measured are located perpendicular to one of these axis lines. All points to be measured are then related to both the Y-axis line and the X-axis line. The measuring technique is simple to conduct and easy to recreate at a later time, such as on a drawing of the accident scene. The measuring technique would be analogous to laying a grid on top of the entire accident scene to allow a means of identifying and quantifying where all the evidence and highway features are located.

Establish a Reference Point and Reference Lines

At an intersection, you can use the curb lines, pavement edge lines, or highway edge lines to establish reference lines and a Reference Point (RP). Typically, an imaginary extension of that line or road edge is made across the intersection; this is sometimes referred to as a lateral prolongation or lateral extension. The pavement edges or curb lines would then be the reference lines.

If you are not comfortable visualizing an imaginary line, then you can stretch a long tape measure (a 100'-long tape measure works well) at the curb line or pavement line edge. If it is safe to do so, you can stretch the tape measure across the intersection so that the tape is in alignment with the curb lines or pavement edges on both sides of the intersection. A string line stretched along the curb line or the pavement edge also works well. You then repeat the process for the second set of curb lines or pavement edges present at the intersection. Do not create a hazard to other motorists when placing tapes and string lines in the traveled portion of a street or highway.

The point where the two lines cross is the RP. This point, which is sometimes called the origin or zero point, is then used as the starting point for all measurements you take at the scene of the accident. Use the base of the curb edge with the pavement for the line location (see Figures 6–4 through 6–5).

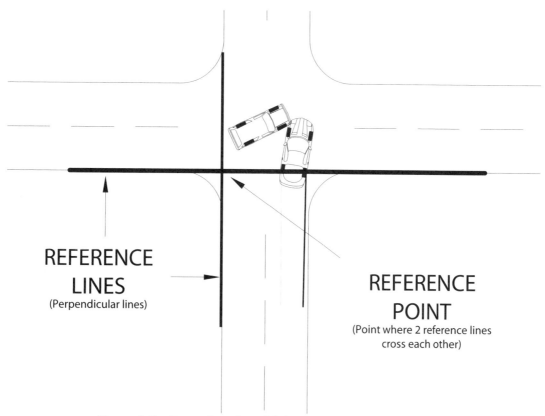

Figure 6–3   Examples of establishing reference lines and RPs.

64   Recording Techniques

Figure 6–5 illustrates the process of creating reference lines at each road edge and establishing an RP where those two lines intersect.

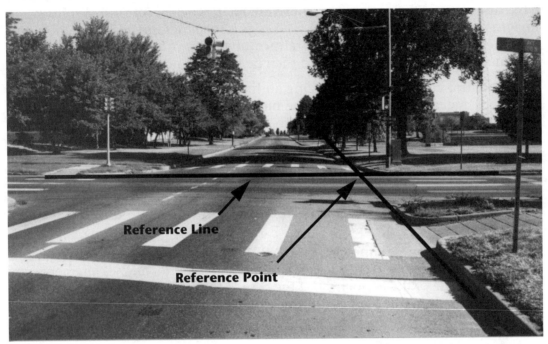

Figure 6–4   The arrow points to the RP created by the junction of the prolongation of the two curb lines as depicted by the two lines.

Figure 6–5   Reference point and reference lines.

When there is no intersection, use the features of the highway to establish reference lines and an RP (see Figure 6–6).

In Figure 6–6, an imaginary line was established across the roadway at a point opposite a speed limit sign. It is helpful if the temporary RP you have chosen is also located with respect to some other permanent fixture, such as a bridge structure, a mile post, or an intersecting street.

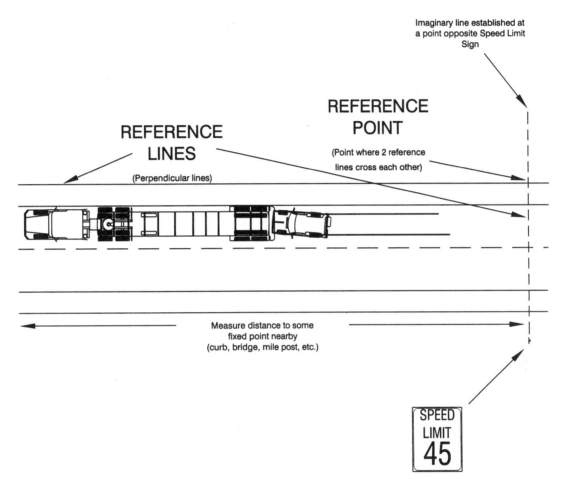

Figure 6–6 Establishing reference lines.

If the temporary reference line cannot be located at a later time (the speed limit sign in the example may be moved or knocked down), the permanent reference can be used as a backup. This process allows you or someone else to return to the same location at a later time and obtain additional or corroborating measurements by identifying and perhaps utilizing the same initial RP that you selected at the time of your initial investigation.

Figure 6–7 illustrates another example of a temporary RP by using a pavement edge line and a traffic control sign as reference lines.

66 Recording Techniques

Figure 6–7 Temporary RP created using a pavement edge line and extension from a traffic control sign. *Photo courtesy of Colorado State Patrol.*

If there is no highway fixture at the accident site (like the speed limit sign), you can establish an RP arbitrarily along the road edge, lane line edge, or other suitable object. The temporary established point should be located and measured with respect to a permanent fixture, as previously described (see Figure 6–8).

Figure 6–8 Temporary RP.

Use the same methods on a curved road as you would on a highway or mid-block section away from an intersection. Use one edge of the road as a reference line and another perpendicular reference line to establish an RP (see Figure 6–9).

In Figure 6–9, an RP was established along the road edge at a position opposite the end of the guard rail. Measurements can be taken along the road edge (east–west) and across the road (north–south) from the RP.

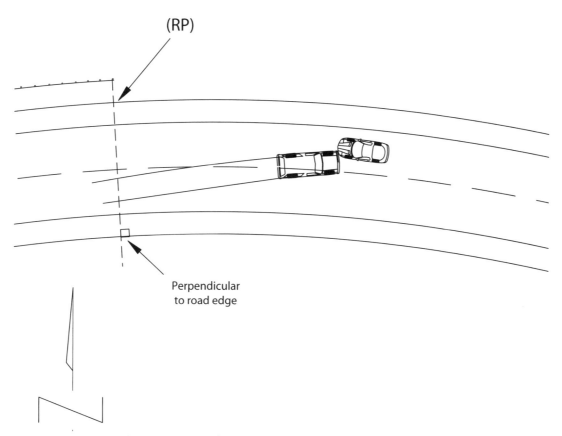

Figure 6–9 Method to determine RP on a curved road.

Decide on the points to be measured and include the measurements in the field sketch. For each point, you will need two measurements. One measurement will be the north-south dimension from your first reference line (X-axis), and the second measurement will be the east-west dimension from your second reference line (Y-axis). All of your measurements will start at the RP (see Figure 6–10).

The examples of measuring techniques shown in Figures 6–11 and 6–12 illustrate the various points or locations that you may need to document at an accident scene. Note on both diagrams that the wheel/axle positions of each vehicle are locations that were selected for documentation. This process establishes the vehicles' positions in the roadway, either at rest or during the creation of evasive action tire marks prior to the crash.

In your diagram, include a table for recording the measurements (see Figure 6–13). Take the measurements along the road edge (east—west in the example) and across the road (north–south in the example) from the RP. Remember that for each point measured, you'll need two measurements (one measurement in the east–west direction and one measurement in the north–south direction). You may also want to include a description of each point measured.

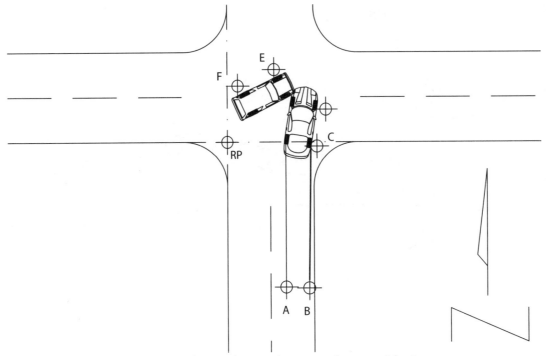

Figure 6–10 Various points to document in an accident scene.

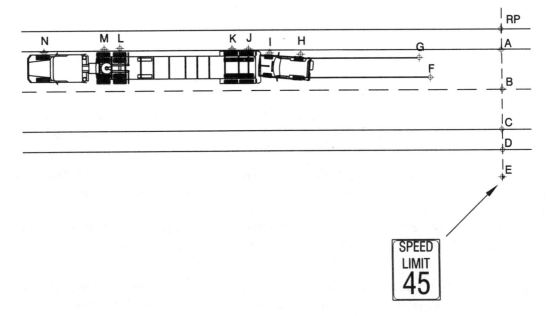

Figure 6–11 Points to be located on a rear-end collision.

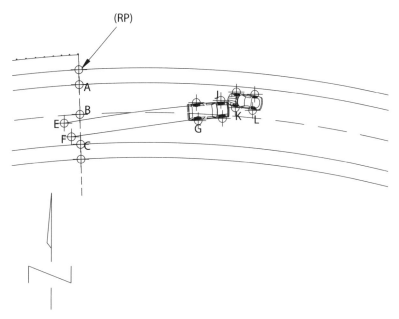

Figure 6–12 Points to be located for a head-on crash on a curve.

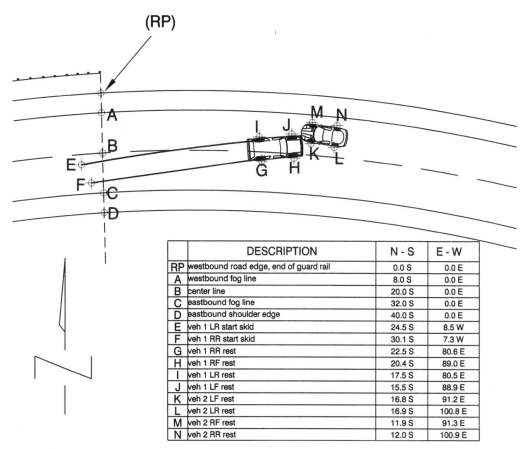

	DESCRIPTION	N - S	E - W
RP	westbound road edge, end of guard rail	0.0 S	0.0 E
A	westbound fog line	8.0 S	0.0 E
B	center line	20.0 S	0.0 E
C	eastbound fog line	32.0 S	0.0 E
D	eastbound shoulder edge	40.0 S	0.0 E
E	veh 1 LR start skid	24.5 S	8.5 W
F	veh 1 RR start skid	30.1 S	7.3 W
G	veh 1 RR rest	22.5 S	80.6 E
H	veh 1 RF rest	20.4 S	89.0 E
I	veh 1 LR rest	17.5 S	80.5 E
J	veh 1 LF rest	15.5 S	88.9 E
K	veh 2 LF rest	16.8 S	91.2 E
L	veh 2 LR rest	16.9 S	100.8 E
M	veh 2 RF rest	11.9 S	91.3 E
N	veh 2 RR rest	12.0 S	100.9 E

Figure 6–13 Use of a table to record measurements and point descriptions.

Triangulation Measurement System

Triangulation measurements differ from coordinate measurements in that every point is located with respect to two RPs. As the name implies, a series of three measurements is made for each point identified. RPs in the triangulation measurement system can be locations similar to those in the coordinate measurement system (extensions of curb lines, intersecting lines, etc.) or they can be fixed objects (utility poles, bridge rail ends, identified fence posts, etc.).

You must take three measurements for each point located: the distance from RP1 to the point; the distance from RP2 to the point; and the distance from RP1 to RP2 (see Figure 6–14).

The point located in Figure 6–14 is measured with respect to two RPs (in this case, two utility poles). The distance between the two utility poles, or RP1 and RP2, would be the same if additional locations are documented by triangulation measurements. When recording your measurements, be certain to record the general compass direction of each measured point relative to the two established RPs. For example, if you are measuring to the left front wheel of the car that created the skid marks, one measurement would be 22 feet southeast of RP1. The second measurement would be 19 feet southwest of RP2.

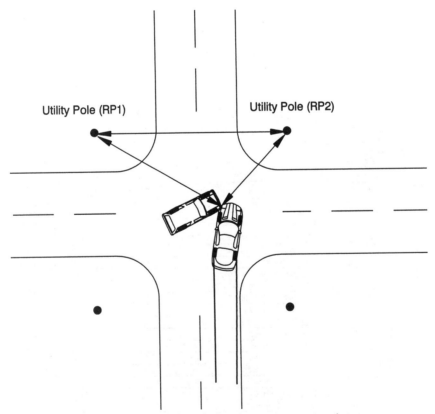

Figure 6–14 Triangulation measurement points.

The same measuring protocols apply when using the triangulation method. Every point is measured from at least two fixed RPs. To establish the curvature in the skid marks created by the truck tractor and semi-trailer in Figure 6–15, you would need to measure to at least three points along the curve path of the tire marks.

Figure 6–15   Measuring protocols when using triangulation method.

Triangulation is used infrequently by investigators; it is used when there is difficulty in setting up and using a coordinate measuring system. Typically, triangulation is used when there are irregular road edges, when the items to be measured are a significant distance from the highway, or when the road is made up of substantial curves (like a narrow, winding mountain road). Triangulation measuring techniques may also be used if the accident scene data will be developed into a scaled map or diagram.

Figure 6–16 shows where large pieces of debris and the final position of rest of a pickup truck that traveled off of the downhill grade of the highway were located. You may document this more accurately by utilizing the triangulation method.

72 Recording Techniques

Figure 6–16 Triangulation method needed to accurately document accident scene evidence that is spread out.

Triangulation Method of Measuring a Curved Road

Figure 6–17 and Figure 6–18 show triangulation measurements on a curved road.

Marking Materials

You might consider using various marking materials to assist you in locating and measuring information on a street or highway. As mentioned in the Tools, Equipment and Supplies section in Chapter 2, you can use materials to mark the road surface. These materials include the following:

- Sidewalk chalk
- Waxed crayons
- Spray paint
- Liquid white shoe polish

You may place these materials on the road surface temporarily. They are most visible on a dry road surface. Be careful to use only small quantities of spray paint, as its use may be deemed to deface a public highway by some Departments of Transportation.

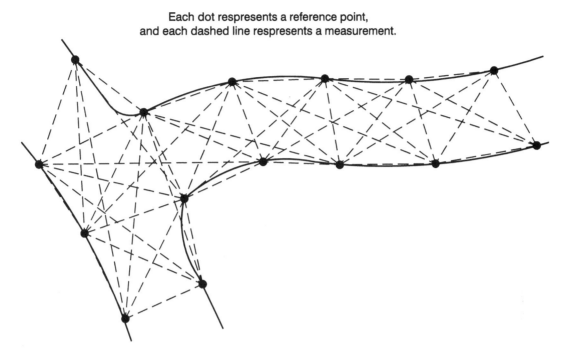

Figure 6–17 Each dot represents a reference point and each dashed line represents a measurement.

Figure 6–18 The curvature of the road at this accident scene would indicate that triangulation measuring techniques should be used.

74 Recording Techniques

In the photographs shown in Figure 6–19, an arrow and a letter were created using all of the materials listed. The surface in the left photograph is concrete. The surface in the right photograph is asphalt.

You can photocopy the chart in Figure 6–21 and utilize it with either coordinate or triangulation measuring techniques.

Figure 6–19   Arrows on concrete (left) and asphalt (right).

Figure 6–20 In this photograph the officer has used spray paint to identify various gouges and tire marks within the area of impact at an intersection location.

Traffic Accident Measurement Record

File or Case #_____ **Location**_____

Date of Accident_____ **Date of Measurements**_____ **Time**_____

Measurements By_____ **Reference Point Location**_____

#	Description of Measurement or Location	Direction	Direction

Figure 6–21 Traffic accident measurement record form. This chart can be photocopied and utilized with either coordinate or triangulation measuring techniques.

Accident Templates

You can use different accident templates in your investigation of a traffic accident to draw the various symbols shown in Figure 6–22. These symbols can be incorporated into your field sketch, as appropriate.

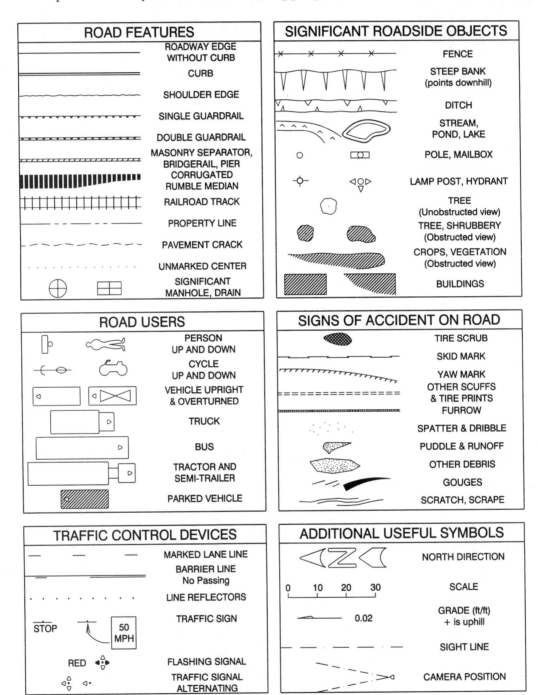

Figure 6–22 Recommended symbols for accident-location maps.

The template in Figure 6–23 is produced and marketed by the Traffic Institute, Center for Public Safety, Northwestern University.

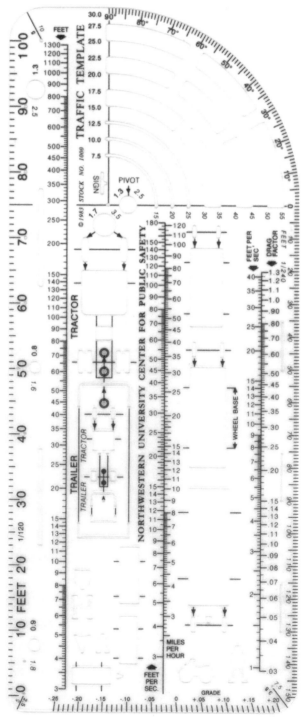

Figure 6–23 Template from Traffic Institute, Center for Public Safety, Northwestern University, Evanston, Illinois.

78 Recording Techniques

The template in Figure 6–24 is produced and marketed by the Institute of Police Traffic Management, University of North Florida, and is specifically for use with commercial motor vehicles.

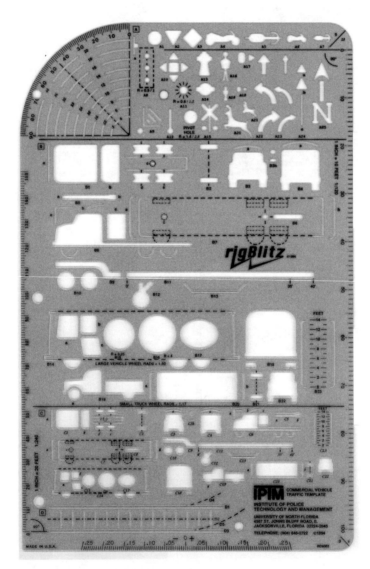

Figure 6–24 Template from Institute of Police Traffic Management, University of North Florida.

7

Photographic Techniques

The information you can document through the use of photographic equipment can be substantial. The quality of that information may be improved by understanding how the camera operates, what its limitations are, what its features will allow you to do, and what skills you have. This chapter will not focus on which camera is the best one for you, as that is an individual choice dependent on many factors. There are some techniques that you can employ, however, that will improve the quality of your results and the value that your photograph will have in assisting you and others as you analyze and evaluate the crash.

Vehicle Examinations

Start with the "basic four" photographs of the vehicle: a front view, rear view, and the two side views (see Figure 7–1). Take your photographic shot perpendicular to the side of the car or truck you are documenting.

This process allows you to obtain an overall view of each side of the vehicle and to show perspective on the extent of physical damage to the vehicle. It also allows you to determine at a later time the approximate dimensions of the

80 *Photographic Techniques*

Figure 7–1 The basic four photos—front, back, and sides.

damage location and the extent of collapse on the vehicle, if needed. Keep in mind that others may want to examine the photos and will not have the benefit of viewing the vehicle in person, as you did. You can photograph large truck and motor coach units in the same manner.

The approximate camera positions for the basic sequence of four photographs (see Figure 7–2) is shown using the solid arrows. The dashed arrows illustrate camera positions for supplemental exterior photographs.

By using this technique, you will be able to capture the overall view of each of the four sides of a vehicle, as well as corner views of the vehicle showing two adjacent sides of the vehicle concurrently. You can use this technique with large commercial vehicles also.

You can obtain interior photographs of the vehicle by utilizing this same pattern. You should take these photographs much closer to the vehicle, so that details of the appearance, condition, and damage within the passenger compartment may be documented.

If possible, attempt to photograph the vehicle when it is not crowded in next to other vehicles. Many times this is not possible due to circumstances beyond your control. If the vehicle cannot be moved safely or efficiently, as shown in Figure 7–3, then take the photos as best you can.

Vehicle Examinations 81

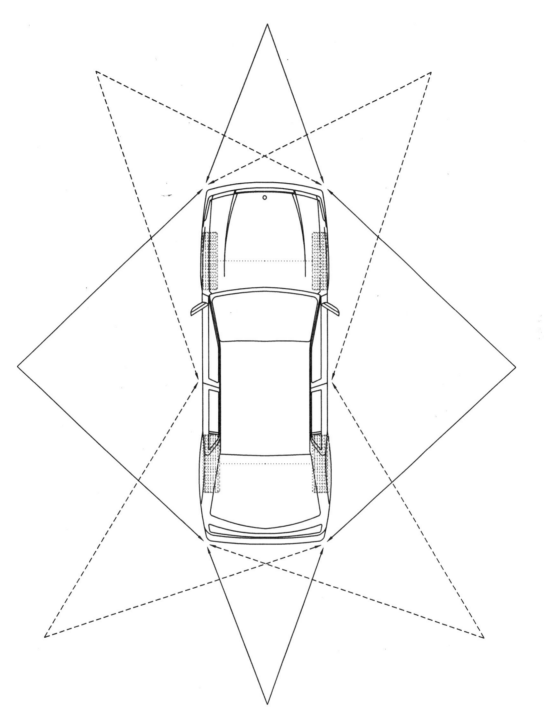

Figure 7–2   Camera position for basic and supplemental photos.

As strange as it may seem, use the electronic flash on your camera when taking any photographs of a vehicle. The flash in many still cameras will suffice for most daytime photos of the vehicle. This is especially important on sunny days, when dark shadows may exist outside, underneath, and inside the vehicle. The

82 Photographic Techniques

Figure 7–3   Photos of vehicles that could not be moved to be photographed.

flash will provide supplemental illumination and "fill in" the darker areas of the vehicle that are covered with the shadow. This will create a better photograph. As shown in Figure 7–4, areas in the right rear wheel well and the interior of the vehicle probably would not have been illuminated adequately if a flash unit had not been utilized when the photograph was taken.

Figure 7–4   Using flash to fill in dark areas.

If you are taking photographs of the vehicle at night, you may want to utilize a larger supplemental flash with your camera. If you are taking photographs in low-light conditions, such as outside at night or in an unlighted garage, you may also want to consider "bracketing" your shots. This process changes the exposure settings on your camera to increase the aperture setting (the amount of lens opening that allows light to enter the camera) or the shutter speed (changing the amount of time that the shutter remains open). This increases the potential that you will obtain a properly exposed photograph.

When taking photographs, consider utilizing a measuring device in the photo so that scale, dimension, and perspective can be determined easily. The measuring device can be a carpenter's tape measure, a yard stick, or a surveyor's rod. Place the measuring device vertically or horizontally within the area encompassed by the photograph. When you take the photograph, be certain to place the camera at a level so that the photograph will be perpendicular, in the vertical alignment, to the subject area of the photo. The two photos at the top of Figure 7–5 illustrate a downward angle of the camera to the vehicle, which diminishes the accuracy of any subsequent measurements from the photographs.

Figure 7–5   Improperly photographing measuring devices at an angle (top) and properly using them with perpendicular vertical alignment (bottom).

Take photographs of the entire exterior of the vehicle, following a perimeter pattern around the outside of the vehicle. Take these photos at different heights, so that the entire exterior surface of the vehicle is photographed. Remember, that you are documenting surfaces of the vehicle that exhibit damage from the crash, as well as surfaces of the vehicle that were not damaged.

By following a pattern around the vehicle, you will be certain to capture all of that evidence. On some vehicles, you may consider taking perimeter photographs of the lower portion of the vehicle and then perimeter photographs of the upper perimeter of the vehicle. By using this process, as depicted in Figure 7–6 and Figure 7–7, with sport utility vehicles, pick-up trucks, and larger trucks and motor coaches, you will be able to show sufficient detail of the appearance and post-accident condition of the vehicle.

Approximately four to six photographs on both the right and left sides of the vehicle should suffice for a passenger car, sport utility vehicle, or pickup truck. Two to three photographs across the front and rear may also be helpful to document the condition and appearance of those areas of the vehicle. More photographs may be needed on a larger commercial motor vehicle. Once you have finished that process, you can the focus on close-up photographs of the damaged areas on the vehicle.

Figure 7–6 Upper perimeter photographs and lower perimeter photographs.

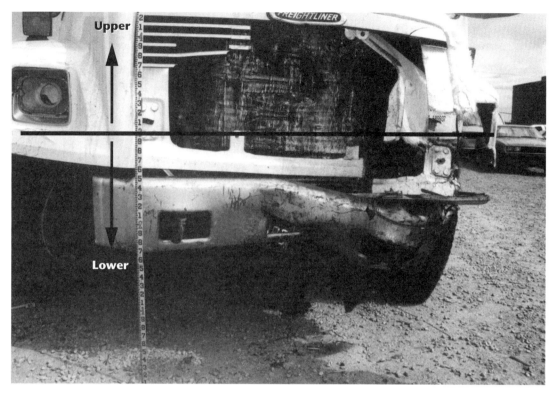

Figure 7–7   Upper perimeter photographs and lower perimeter photographs.

Your measuring device can also be used to demonstrate the extent of collapse of a vehicle as a result of a collision. In Figure 7–8, the pick-up truck collided with the right rear corner of a flatbed semi-trailer. The speed differential was approximately 40 miles per hour. The end of the measuring device, as shown by the arrow, was placed to depict where the position of the left-side forward roof support pillar, also known as the "A" pillar, was positioned prior to the collision.

Vehicle photographs should also be taken to illustrate the shifting of cargo on a trailer and the presence or absence of safety-related items on a vehicle (see Figure 7–9).

86 Photographic Techniques

Figure 7–8 Using a measuring device to show extent of collapse.

Figure 7–9 Image showing cargo shift and absence of safety-related items.

Highway Photographs

When taking highway photographs, attempt to place the camera parallel with the travel lane(s) that you are documenting. Obviously, your safety is of primary concern, so do not attempt to enter the travel lanes of a highway unless it is safe. Your time in a traffic lane easily could be five to 10 seconds or more, so traffic within 1,000 feet may be an imminent hazard to you. The approaching driver may also not recognize your presence or respond as you anticipate. If you have the opportunity to photograph an accident scene with the highway closed by authorities, take advantage of the situation.

The six photographs in Figure 7–10 depict approximately 100 feet of the approach path of one driver to a head-on crash. They allow you to place the impact area relative to the traffic lane and to establish the lack of pre-impact evasive action.

Figure 7–10   Sequence shows 100' path of approach of one driver in a head-on crash.

Your scene photographs should also show details relating to how the traffic crash occurred and the relationship of evidence to final rest positions. In Figure 7–11, the tire mark (as noted by the arrows) suggesting overdeflection of the right-side tires, identified the path of the truck tractor and semi-trailer and indicated that the speed of the truck was not significant.

Figure 7–11   Relationship of evidence to final rest position. Tire marks suggest overdeflection.

As with vehicle photographs, utilize a measuring device, as shown in Figure 7–12, to illustrate the dimensions, perspective, and position of physical evidence, such as tire marks and pavement gouges, that are present on the highway surface. Take the photographs parallel with the travel lane to allow a proper perspective of where the evidence is located within the highway lane or street.

Take photographs and/or video of the approach path that each driver utilized while traveling towards the crash site. The sequential documentation may be taken through the windshield of a moving vehicle or you can walk back the distance and then approach the accident scene on foot while taking photographs. If the accident occurred at an intersection, then take a series of photographs on each respective approach street or highway.

When documenting the route utilized by a driver, photographically record a time duration of at least five seconds before the crash event. The distance from the accident location where you should start your photographs is shown in Table 7–1. The distances listed in the chart should be considered minimums. You may want to start your series of photographs at a greater distance from the area of impact,

Figure 7–12 Use of measuring device with highway evidence.

Table 7–1: Minimum Distance Photography Chart

Posted Speed Limit	Equivalent Velocity	Distance Traveled in 5 Seconds
30 m.p.h.	44 feet per second	215 feet
35 m.p.h.	51 feet per second	255 feet
40 m.p.h.	58 feet per second	295 feet
50 m.p.h.	73 feet per second	365 feet
55 m.p.h.	80 feet per second	400 feet
60 m.p.h.	88 feet per second	440 feet
65 m.p.h.	95 feet per second	476 feet
70 m.p.h.	102 feet per second	510 feet
75 m.p.h.	110 feet per second	550 feet
80 m.p.h.	118 feet per second	586 feet

as listed in Table 7–1, if there are particular highway features, topographical variations, or commercial developments along the side of the highway that you would like to document.

If possible, you should attempt to take these photographs while traveling in the same lane of travel as the accident vehicle. That way, you will document the approximate view that the driver had during their approach to the accident scene. If the sunlight angle was a factor, then you may want to consider taking the photograph series at a similar time of day to capture its effect on a driver.

The sequence of photographs, shown in Figure 7–13, would allow you to document the appearance, condition, lane configuration, and alignment of the approach path taken by a driver. You will be also to illustrate the line-of-sight limitations, if any, that a driver may have had while approaching the location where the collision eventually occurred.

Figure 7–13 Photo sequence documents path traveled by a driver.

You may photocopy the following form to assist you with recording information regarding the photographs taken.

File or Case #_____ **Subject of Photos**_____ **Roll #**____
Date of Accident_____ **Time of Accident**_____ **Location**_____
Date of Photos_____ **Time of Photos**_____ **Taken By**_____
Camera Position _____ **Film**_____ **Flash** ❏

	Direction	Description of Photo
1		
2		
3		
4		
5		
6		
7		
8		
9		
10		
11		
12		
13		
14		
15		
16		
17		
18		
19		
20		
21		
22		
23		
24		
25		
26		
27		
28		
29		
30		
31		
32		
33		
34		
35		
36		

8

Vehicle Information

Information regarding the vehicle(s) involved in a traffic accident is one of the three important areas (human, vehicle, and environment) of traffic accident investigation. The vehicle provides information and evidence that will allow you to correlate physical evidence created on the pavement surface with the specific vehicle that created the evidence. Information from the vehicle also gives you the ability to evaluate the motion and dynamics relevant to the traffic accident.

The information you need to gather from any vehicle involved in the crash, whether a passenger car, motorcycle, sport-utility vehicle, pick-up truck, or commercial motor vehicle, can be broken into the following three distinct areas:

- Basic vehicle identification, dimensional data, loading and background information
- Evidence of contact and/or physical damage resulting from the collision event
- Evidence of the lack of contact or physical damage and the overall condition of the vehicle

A systematic process of collecting and documenting this information will allow you to understand and evaluate the traffic accident better. Vehicle background information will allow you to conduct research regarding both your vehicle and any other vehicles involved in the accident. For example, it may be of importance that one of the vehicles involved in the crash had an anti-lock brake system, an automatic headlight and tail lamp illumination system, or cruise control. Information regarding climate control settings may indicate if the defroster system was operating at the time of the crash.

You could integrate the background information on a vehicle with documentation regarding the load in the vehicle at the time of the accident. This information may indicate that the additional weight that was carried in the vehicle at the time of the collision exceeded the gross vehicle weight rating. This factor may affect the vehicle's ability to swerve or stop. The background information on a vehicle may indicate the type of seat belt restraint installed in the vehicle. An examination of the seat belt restraint system may indicate whether a person was wearing the seat belt restraint installed in the vehicle at the time of the crash. If you know when the vehicle was manufactured, you can also research all of the manufacturer safety bulletins and mechanical recalls for the year, make, and model of the vehicle in question.

Table 8–1 includes a more detailed review of information you may want to consider documenting when examining a vehicle.

Table 8–1: Vehicle Examination Data Gathering

Vehicle Data	Year, make, model, trim package
Vehicle Identification	Vehicle Identification Number (VIN), license plate registration, mfg. date
Engine	Gasoline, diesel, CNG or electric; engine size, number of cylinders; HP rating
Drive Train	Automatic transmission (3-, 4-, 5-speed), Standard transmission (3-, 4-, 5-speed); 2WD, 4WD, AWD, FWD, RWD, OD
Body Style—Passenger Vehicle	2-door, 3-door, 4-door; sedan, coupe, convertible, wagon, lift-hatchback; sliding door, SUV, MPV, passenger van
Body Style—Commercial Van	Slide side doors, panel doors—side, panel doors—rear; passenger capacity
Body Style—Pick-up Truck	2-door, 3-door, 4-door; ½ T, ¾ T, 1 ton reg. cab, extend. cab, crew cab; 2WD, 4WD; fleet side bed, step side bed, short bed, long bed, flat bed, utility bed

By recording this information, you can then research background materials and manufacturer specifications on a given vehicle. This will allow you to collect information on whether the vehicle has been involved in a previous crash and then salvaged, as well as ownership records of the vehicle through a title search. You can search maintenance records, if needed, and compare recall data on the make and model to maintenance activities and repairs. You can also research manufacturer specifications for a given year, make, and model of some vehicles to obtain acceleration capabilities, stopping distances, turning radius data, and tip-over potentials.

The Vehicle Identification Number (VIN) is probably the most important piece of data to collect when examining a vehicle. The VIN is a unique combination of alphanumeric characters assigned to a specific vehicle, formulated by each manufacturer and presented in a format required by federal standards. All vehicles manufactured after 1981 will have a total of 17 characters in the VIN (see Figure 8–1). The VIN will lead you to a great deal of information regarding the vehicle. See Figure 8–2 for location of VINs.

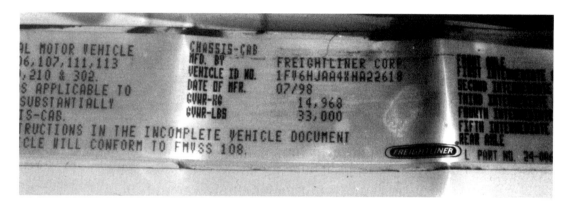

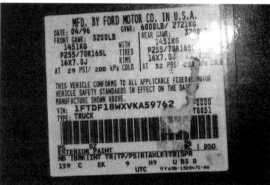

Figure 8–1   The Vehicle Identification Number (VIN).

Many vehicles are equipped with after-market items—accessories that have been installed on the vehicle at some time by someone other than the original equipment manufacturer. It is likely that some of these features, accessories, and convenience items may have been installed on a vehicle involved in the accident by an after-market company or vehicle owner. You should note this information, as after-market additions to a vehicle may affect the vehicle operations, such as the visibility potential for a driver, handling characteristics of the vehicle during a certain type of maneuver, and/or the weight distribution on a given axle (see Figure 8–3).

96 Vehicle Information

Figure 8–2 Typical VIN plate locations on passenger cars, sport-utility vehicles, and pick-up trucks are at the left door frame and the base of windshield on left side.

When examining the vehicle, also look for evidence of prior physical damage that has been repaired. Evidence of automotive body putty, different paint colors on vehicle components, and improperly fitting exterior components and parts

Figure 8–3 Note after-market items.

might indicate a previous collision event. This previous event may have affected the vehicle's structural integrity to withstand collision forces and may be indicative of the presence of unrepaired or improperly repaired physical damage on a vehicle.

The list in Table 8–2 illustrates some of the components and safety items that you may want to examine and document during your vehicle inspection. You may duplicate this list for use during a vehicle examination.

Additional information regarding the vehicle may also be helpful to you as you investigate and analyze the traffic accident. Background information regarding dimensional data for the vehicle can be helpful in your investigation, as it will allow you to compare measurements of tire mark evidence with a particular vehicle. It will also allow you to compare the amount of crush damage on a vehicle with the dimensions of an undamaged exemplar vehicle. By utilizing a vehicle schematic similar to Figure 8–4, you can record longitudinal, lateral, and vertical measurements of damage locations and graphically depict patterns of physical damage on the vehicle's exterior surface.

The process of documenting data and damage to the vehicle involves a systematic recording of information related to each of the vehicles involved in the traffic crash. This information can be generated from several sources, the most likely being a personal inspection of all of the vehicles involved in the crash.

Table 8–2: Vehicle Components and Safety Items

Power Options	**Décor and Comfort**
Power Steering	Air Conditioning
Power Brakes	Dual A/C
Power Windows	Electric Rear Defogger
Power Locks	Tilt Steering Wheel
Driver Seat (__way)	Cruise Control
Pass Seat (__way)	Cloth Seats
Power Antenna	Vinyl Seats
Power Left Rear-view Mirror	Leather Seats
Power Right Rear-view Mirror	Bench Seats
Power Trunk Release	Bucket Seats
	Rear Wiper
Wheels	Intermittent Wipers
Standard Wheel Cover	Body Side Molding
Custom Wheel Cover	Center Console
Spoke Wheel Cover	Reclining Front Seats
Aluminum/Alloy Wheels	Keyless Entry
Oversize Tires	Custom Paint
Low-profile Tire	Custom Interior
Temporary Spare	Wood-grain Interior
Full-size Spare	Heated Seats
Trucks – MPV – SUV	**Trucks – MPV – SUV**
Step Bumper	Bed Slider
Slide Rear Window	Fixed Tool Box
Auxiliary Fuel Tank	Grille Guard
Engine Retarder	Dual Rear Wheels
Running Boards	Utility Shell
Electric Winch	Slide-in Camper
Auxiliary Lighting	Ground-effect Add-ons
Fog Lights	Decals
Bed Liner—Slider	Custom Paint
Chrome Bed Rails	Tailgate Removed
Trailer and Tow Equipment	
Roll Bar	
Sound	**Restraint System**
AM Only	Lap Belt Only (Front Seats)
AM/FM	Lap Belt Only (Rear Seats)
XM Radio	Lap and Shoulder (Front Seats)
Cassette Player	Lap and Shoulder (Rear Seats)
CD Player	Motorized Belts (Front Seats)
CB Radio	Driver Air Bag (Deployed? Yes or No)
Custom Install	Passenger Air Bag (Deployed? Yes or No)
Cell Phone	Side Air Bag (Deployed? Yes or No)
	Child Seat
	Infant Seat

Table 8–2: Vehicle Components and Safety Items (Continued)

Safety Items	Roof Options
ABS Brakes	Hard Top
OnStar System	Soft Top (Convertible)
Daytime Lights	Vinyl Cover
Auto Headlights	Glass T-top
Heads-up Display	Steel T-top
Navigation System	Electric Retractable
Collision Avoidance System	Roof Rack

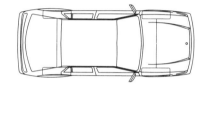

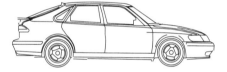

Figure 8–4 Vehicle schematics for recording damage locations and graphically depicting patterns of physical damage on the vehicle's exterior.

A second source of vehicle damage information may come from photographic documentation created by an insurance claims adjuster, a third party insurance administrator, a vehicle damage appraiser, or a law enforcement officer. These photographs may be available for your review.

100   Vehicle Information

A third source of vehicle information could be reports or documents relating to the vehicle. These include damage estimate reports, total loss appraisal reports, and vehicle condition reports generated after the collision by a damage appraiser or claims adjuster.

General vehicle data resources are also useful, such as those found in automotive and vehicle books and magazines marketed to consumers. Other sources include those found on the Internet, in government-related databanks, or through computer software companies. You should also consider sales literature from dealers if you are researching information of a late-model vehicle.

The same process of recording information about a vehicle applies to medium- and heavy-duty commercial vehicles. Although the basic information documented from commercial vehicles is similar to that from passenger cars, sport-utility vehicles, and pick-up trucks, the quantity of information from a commercial vehicle may be greater (see Figure 8–5).

```
Engine Mfg:_____ Model_____ HP_____ ❑ Gas ❑ Diesel ❑ Turbo
Transmission:_____ Model_____ # of Speeds_____ OD_____ Aux Trans_____
Rear Axle(s) Mfg:_____ Model_____ ❑ Single ❑ Tandem ❑ Twin Screw ❑ Tag ❑ Pusher
Frame: ❑ Steel ❑ Aluminum ❑ Extended ❑ Standard Length    Mfg Listed Wheel base _____Inches
Fifth Wheel Mfg: _____ ❑ Fixed ❑ Sliding ❑ Air ❑ Manual ❑ Air Release
Fuel Tank(s): ❑ Mfg. Standard ❑ Custom ❑ Twin Tanks ❑ Polished ❑ Step ❑ Saddle
Front Suspension: Rated Capacity_____ ❑ Leaf ❑ Shock
Rear Suspension: Rated Capacity_____ ❑ Leaf ❑ Air Bag ❑ Torsion ❑ Hendrickson ❑ Spring
Steering: ❑ Manual ❑ Power Assist
Brakes: ❑ Hydraulic ❑ Air ❑ Air Brake Dryer
Exhaust: ❑ Single ❑ Dual ❑ Under ❑ Cab ❑ Vertical ❑ Chrome
Front Axle Wheels: ❑ Disc ❑ Spoke ❑ Aluminum ❑ Polished ❑ Chrome ❑ Painted
Rear Axle Wheels: ❑ Disc ❑ Spoke ❑ Aluminum ❑ Polished ❑ Chrome ❑ Painted
Rear Axle Wheels: ❑ Disc ❑ Spoke ❑ Aluminum ❑ Polished ❑ Chrome ❑ Painted
Left #1 Axle _____     Size_____ Tire Pres_____ Depth_____
Left #2 Axle Outboard_____  Size_____ Tire Pres_____ Depth_____
Left #2 Axle Inboard_____  Size_____ Tire Pres_____ Depth_____
Left #3 Axle Outboard_____  Size_____ Tire Pres_____ Depth_____
Left #3 Axle Inboard_____  Size_____ Tire Pres_____ Depth_____
Right #1 Axle_____   Size_____ Tire Pres_____ Depth_____
Right #2 Axle Inboard_____  Size_____ Tire Pres_____ Depth_____
Right #2 Axle Outboard_____  Size_____ Tire Pres_____ Depth_____
Right #3 Axle Inboard_____  Size_____ Tire Pres_____ Depth_____
Right #3 Axle Outboard_____  Size_____ Tire Pres_____ Depth_____
```

Figure 8–5   Recording information from a commercial vehicle.

Other information pertaining to a commercial vehicle may be helpful to record for your investigation, such as truck accessories and components, especially because commercial motor vehicles are usually not manufactured generically (as Ford Taurus SE passenger sedans are, for example). Instead, they are built specifically for a customer. For example, two Peterbilt Model 379 truck tractors undoubtedly will be equipped differently, even though they were built side-by-side on the assembly line. See Figure 8–6 for components you should consider recording.

❏ Conventional Cab ❏ Cab Over Engine ❏ Day Cab ❏ Straight Truck ❏ Integral Sleeper Berth ❏ Add-on Sleeper ❏ Sleeper Berth Size(____) ❏ Stand-up Berth ❏ Walk-in Berth ❏ Flat-top Berth ❏ High-rise Berth ❏ Add-on Berth ❏ Customized Berth ❏ Air Ride Driver Seat ❏ Air Ride Passenger Seat	❏ Steel Cab ❏ Aluminum Cab ❏ Fiberglass Hood ❏ Steel Hood ❏ Custom Paint ❏ Custom Lettering ❏ Standard Interior ❏ Custom Interior ❏ Classic Interior ❏ Int: Average Condition ❏ Int: Poor Condition ❏ Int: Excellent Condition ❏ Roof Air Deflector ❏ Side Farings ❏ Full Aero Package ❏ On-board Computer	❏ Fog Lamps ❏ Supplemental Headlights ❏ Add-on Lighting ❏ Spot Light ❏ ¼ Fenders ❏ Dual Air Horns ❏ Tilt Steering Wheel ❏ Ext. Sun Visor ❏ Bug Screen ❏ Headache Rack ❏ Plug-in Heater ❏ Jake Brake ❏ Jake 2__4__6__ ❏ CB radio ❏ AM/FM/Cassette ❏ Cell Phone	❏ GPS ❏ Conspicuity Tape ❏ Supplemental Signals ❏ Std. Side Mirrors ❏ Convex Mirrors ❏ Fender Mirrors ❏ Door-top Mirrors ❏ Crossover Mirrors ❏ Moto-mirrors ❏ Warning Beacons ❏ Rear Cab Window ❏ Warning Signs ❏ Oversize Windows ❏ Tinted Glass ❏ In-door Windows ❏ Warning Lights

Figure 8–6 Truck accessories and components.

Damage Sustained in the Traffic Accident

Evaluating what physical damage resulted from a collision involves recording the areas or components of the vehicle that exhibit physical damage and/or evidence of contact. This allows a subsequent evaluation of *how* that damage or contact occurred.

Northwestern University's *Traffic Accident Investigation Manual, Volume 2*, classifies two categories of physical damage a vehicle may sustain in a collision:

Contact Damage Deformation or defacement resulting from direct pressure of another object or surface in a collision

Induced Damage Damage to a vehicle other than from contact, often indicated by crumpling, distortion, bending and breaking

Other research texts refer to the two types of physical damage classifications sustained on a vehicle as "direct" and "indirect," versus contact and induced, respectively. Your ability to identify the type of physical damage on a vehicle correctly will assist you in understanding how the damage was created and how to match the portion or area of the striking object that was involved in the creation of the damage.

A review of photographs of two damaged passenger cars that collided with each other will assist in understanding the two classifications of physical damage. The area between the two lines in Figures 8–7 and 8–8 is *contact* (direct) damage. This was the area of the Cadillac that struck the station wagon in a broadside collision. Damage to the Cadillac in Figure 8–7 on the left front fender, left door, and the engine hood panel would be classified as *induced* (indirect) damage. The damage to Chevrolet station wagon in Figure 8–8 on the tailgate, window glass, and roof panel is induced damage, as it was not directly involved in the surface contact areas of the colliding vehicles.

Figure 8–7   Area between arrows shows contact (direct) damage. Other damage is induced (indirect) damage. *Photo courtesy of Colorado State Patrol.*

Figure 8–8   Area between arrows shows contact (direct) damage. Other damage is induced (indirect) damage. *Photo courtesy of Colorado State Patrol.*

As shown in Figure 8–7, the width of contact damage on the front of the Cadillac is similar in dimension to the width of contact damage on the station wagon. The remainder of physical damage on both cars is classified as induced damage, because that damage was not created directly from contact with another part of the other car. Induced physical damage is caused by the forces in the collision being dissipated throughout the remainder of the vehicle. That vehicle can either be the striking vehicle or the struck vehicle. Figure 8–9 shows a relatively narrow contact damage (consistent with a severe collision with a wooden utility pole) and a large area of induced damage.

Figure 8–9 Here the contact (direct) damage is relatively narrow, consistent with a severe collision with a wooden utility pole. The bending and bowing of the remainder of the car is classified as induced (indirect) damage.

Contact damage can also be dispersed widely. As shown in the photograph of the car in Figure 8–10, the width of the contact damage is equal to the overall width of the car.

By properly evaluating the areas of contact damage on the vehicles involved in the collision, you can determine how the two vehicles contacted each other. In Figure 8–11, a narrow amount of overlap occurred between the front of the passenger car and the left outboard wheel of the semi-trailer. Although the damage on the Buick looks extensive, some of the damage was created by the collapse of the left side structure of the car due to the magnitude of opposing forces encoun-

Figure 8–10 The entire front end of the Honda sports car had contact with the right side of the Suburban that was pulling the small utility trailer.
Photo courtesy of Colorado State Patrol.

tered from striking the left outboard wheel and axle of the loaded semi-trailer. Additional damage was created during extrication of the driver of the Buick by the fire department.

Damage to vehicles will also provide information on the angularity of the collision, as well as suggest the post-crash movements that occurred. By evaluating the damage to establish the impact configuration, you can determine the impact position of the two vehicles relative to each other. This will then help you establish their approach route on the roadway immediately before the crash occurred.

An examination of the Jeep Cherokee in the following photographs (see Figure 8–12) indicates that impact forces created physical damage to the vehicle, so that the left front corner of the Jeep was pushed back longitudinally, almost in a straight line. There was no significant angularity between the two vehicles at the time of contact, based upon the damage pattern existing on the Jeep. The amount of overlap between the Jeep and the other vehicle was approximately one foot, and was aligned with the longitudinal axis (front to rear) of the Jeep.

When recreating the position of the Jeep Cherokee at the time it rear-ended the Ford SUV, the impact configuration would appear as shown in Figure 8–13.

The direction of the force application, sometimes called the thrust angle, results in equal longitudinal force applications on both vehicles. The force application will collapse the vehicle structure, depending upon the quantity of the force and

Figure 8–11 Properly evaluating areas of contact can determine how vehicles contacted each other. *Photos courtesy of Jerry Woods.*

Figure 8–12 Examination of impact configuration can help determine vehicle position at time of collision. *Photos courtesy of Colorado State Patrol.*

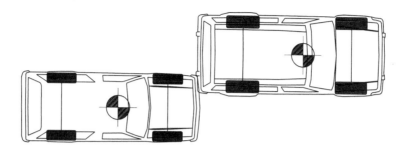

Figure 8–13 Impact configuration between the Jeep and SUV.

the strength of the vehicle structure(s) involved in the contact. The impact configuration shown in Figure 8–14 would result in counterclockwise rotation for both vehicles after impact.

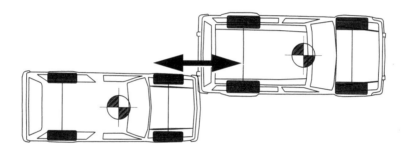

Figure 8–14 Thrust angle.

Using scaled model cars and trucks to evaluate physical damage to vehicles is helpful to correlate how the vehicles collided and what post-collision movements occurred from that impact configuration.

Physical damage does not have to be significant to provide clues as to how the vehicles were configured, or aligned, at the time of impact. The photographs in Figure 8–15 show the front structure and bumper of a truck tractor involved in a collision with a sport-utility vehicle. The damage to the front of the truck tractor was not substantial, but it provided an important piece of evidence.

An imprint of the sidewall of the left rear tire of the sport utility vehicle was created on the front bumper of the truck tractor. The tire imprint has the dimension, as shown by the measuring ruler, of the tire diameter. A close look at the imprint reveals the reverse impression of the shoulder lugs on the tire and the letters "ER", a part of the "WRANGLER" brand name imprinted on the tire sidewall.

108   *Vehicle Information*

This small piece of evidence indicated that the sport utility vehicle was broadside (perpendicular) to the front of the truck tractor at the time of contact. The evidence also indicated that the wheels of the sport-utility vehicle were not rotating, consistent with an aggressive brake application by its driver. This evidence allowed a conclusion that the sport-utility vehicle had spun out of control *prior* to impact by the truck tractor, not as a result of impact by the truck tractor.

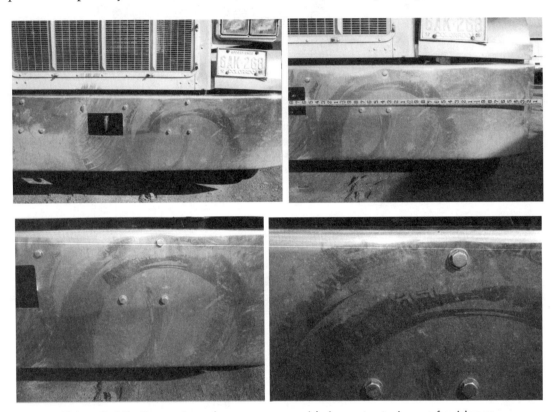

Figure 8–15   Even minor damage can provide important pieces of evidence.

9

Evaluation of the Highway Environment

At some point in your evaluation of a traffic accident, you, or someone else in your company, may be asking the question, "Should I go to the scene?" The answer to that question may depend on the circumstances of the accident, the policies and procedures established in your company, and the location of the traffic crash. It may not be feasible or practical, in some circumstances, to travel to the scene. However, in almost every situation, it is far more beneficial to view the scene of the accident and the surrounding area than to rely on reports. By doing so, you have the opportunity to understand the environment surrounding the accident location, to view traffic patterns and movements, to examine physical evidence on the roadway, to understand information noted in the reports generated by law enforcement officers, and to understand the view perspective of driver and witness statements.

You might ask yourself the following questions as you decide whether to travel to the scene. What tire-mark information will the police measure? How sharp is the entry ramp curve? What physical evidence, that was created before, during, or after the crash will *not* be recorded by the police or state patrol? What circumstances regarding the road surface or alignment might be factors that contributed to the cause of the collision? Should I guess about how far you can see around

Figure 9–1   Accident scene. *Photo courtesy of Colorado State Patrol.*

the curve? Will I be able to get out onto the travel lanes to take measurements and photographs? Perhaps the question of "Should I go to the accident scene?" should be replaced with the question, "What am I going to lose if I don't go to the accident scene?"

Highway Information

Evaluating the highway environment includes locating and documenting highway conditions and contributing factors which led to the occurrence of the traffic crash, determining the highway profile and alignment, noting potential view obstructions, and determining the presence and effect of weather and lighting conditions on and surrounding the highway and the crash location. The sooner you can respond to the accident scene, the greater the opportunity you will have to record and document the conditions and circumstances regarding the accident that were present *at the time* of the accident.

When examining information at the scene of a traffic crash, keep in mind that the "life span" of different types of information and physical evidence is difficult to predict. Table 9–1 reviews the types of evidence and data, including a potential time duration that each type will last.

The nature of the physical evidence and the time duration that it will be evident or visible at the scene is a function of several conditions. These conditions include the following:

Table 9–1: Life Span of Information

Type	Time Duration	Examples
Temporary	From a few minutes up to a few hours; maybe up to a couple of days.	Tire marks on wet surfaces Fluid debris Crash debris Weather/lighting conditions View obstruction/limitations
Short-lived	From a few hours up to a few weeks; perhaps a month.	Tire marks on dry surfaces Fluid debris stains Minor scratching Construction at the scene
Long-term	From several weeks up to a period of years	Deep scratches in the surface Gouges from collision Road configuration/condition Vegetation along the highway

- The size and composition of an evidentiary item or object
- The effects of ambient weather conditions
- The exposure of the roadway evidence to other traffic traversing the area where the evidence is located
- How the evidence was created
- The composition of the road surface
- The surface conditions present in the collision area at the time the evidentiary items were created
- The magnitude of the forces present during the collision that were expressed on the highway or street surface

Figures 9–2 and 9–3 illustrate some of the changes which can occur in the physical evidence created during a traffic collision.

Your evaluation of the highway will focus on several general subject areas. Those areas may allow, create, or affect circumstances that may have contributed to the occurrence of a traffic crash. You should document the existence of these circumstances, as well as their lack thereof, during your investigation of the accident scene.

Investigating visibility limitations at the accident scene is important to understanding what a driver, passenger, or witness may have seen, should have seen, or was unable to see regarding the impending crash event. Visibility limitations can be divided into two categories:

View Obstruction Where vision is completely hindered due to the presence of a solid or nonopaque object. Examples of this type

112 *Evaluation of the Highway Environment*

Figure 9–2 Photograph taken by police officers several hours after the nighttime collisions. *Photo courtesy of Arizona Department of Public Safety.*

of visibility limitation would include buildings, retaining walls, solid fences, and dense foliage

View Interference Where vision is somewhat restricted or limited due to the presence of an object or condition that prevents clear visibility. Examples of this type include small trees, bushes, large profile vehicles, and windshields of parked or moving vehicles

These conditions can occur in several positions; those occurring within the highway, those occurring adjacent to the highway, those occurring off the highway, and those present due to the highway design or alignment.

There also may be visibility limitations due to variations created by numerous atmospheric conditions. These conditions would include:

- Darkness, with or without artificial light sources
- Wind
- Fog
- Rain or mist
- Snow or sleet

Figure 9-3   Photograph of the same area taken by the author three weeks after the crash.

- Glare from the sun or reflected from windows of nearby buildings
- Dust or "liquid splash back" from vehicle travel

Your at-scene evaluation should also include notations or documentation of the highway surface conditions that were present at the time of accident. This knowledge may come from your observations at the accident scene shortly after the occurrence of the crash; from driver, passenger, or witness information; or from evidence noted and documented at the time of the crash. Examples of what you might document include:

- Reduced surface friction
- Evidence of surface treatments (sand, liquid de-icers, chip-and-seal pavement coatings)
- Pavement surface irregularities
- Inadequate shoulder width or soft materials
- Inadequate recovery area or clear zone adjacent to highway
- Pavement drop-offs
- Excessive rutting or pavement surface wear
- Evidence of highway maintenance (plowing, grading, sweeping)

114 Evaluation of the Highway Environment

Incorporated into your evaluation of the highway environment should be data regarding the highway itself. This information may include:

- The composition of the surface materials
- The alignment of the road
- The horizontal curvature (the amount of curvature of the highway towards the right or left)
- The vertical curvature (the amount of uphill and/or downhill grade of the highway)
- The cross-slope profile or the combination of horizontal and vertical curvature (such as a banked, downhill curve)
- The presence of highway fixtures, structures, and traffic control devices (defined as signs, signals, and pavement markings)

A review of accident scenes will assist you in determining what information, at a minimum, should be recorded from different highway situations (see Figures 9–4 through 9–9).

Data to Evaluate for Figures 9–4 and 9–5

• Highway identification	• Number of restricted lanes (trucks, turning lanes, etc.)	• Warning-caution signs
• City, county, and state location		• Flashing warning lights
		• Directional signs
• Compass alignment of road	• Surface abnormalities	• Information guidance signs
	• Shoulder materials	
• Widths of traffic lanes	• Shoulder condition	• Traffic volume
• Number of restricted lanes (trucks, turning lanes, etc.)	• Shoulder warnings (rumble)	• Land use surrounding site
		• Location of accident evidence
• Speed limit	• Shoulder widths (both sides)	• Lane position—accident evidence
• Presence of interchanges	• Roadside vegetation	
• Design of interchange	• Pavement markings	• Dimension of accident evidence
• Surface material and condition	• Condition of pavement markings	• Visibility limitations
• Surface grade and cross slope	• Visibility of pavement markings	• Evidence of weather conditions
• Widths of traffic lanes	• Regulatory signs	• Location of evidence from the nearby interchange

Figure 9–4 Overall view of accident area. *Photo courtesy of David Lohf.*

Figure 9–5 Arrow shows impact evidence. *Photo courtesy of David Lohf.*

116 Evaluation of the Highway Environment

This accident occurred on a two-lane, two-direction, undivided state highway in a rural area. It involved a legal passing maneuver by the truck before the car began to turn left into a gravel pull-out area.

Data to Evaluate for Figures 9–6 and 9–7

- Highway identification
- City, county, and state location
- Compass alignment of road
- Direction of travel of vehicle
- Number of traffic lanes
- Widths of traffic lanes
- Location of Intersecting roads
- Location of intersecting driveways
- Surface material and condition
- Surface grade and cross slope
- Surface abnormalities
- Shoulder materials
- Shoulder condition
- Shoulder widths (both sides)
- Roadside vegetation
- Pavement markings
- Condition of pavement markings
- Visibility of pavement markings
- Regulatory signs
- Warning-caution signs
- Directional signs
- Information guidance signs
- Traffic volume
- Land use surrounding site
- Location of accident evidence
- Lane position—accident evidence
- Dimension of accident evidence
- Visibility limitations
- Evidence of weather conditions
- Location of evidence from interchange
- Post-impact path of travel of truck tractor and semi-trailer
- Post-impact path of red car
- Length of passing zone
- Overhead street light locations
- Speed limit, speed advisories
- Design of intersection
- Curb and gutter compositions
- Curb and gutter style
- Sidewalk presence
- Sidewalk placement (both sides of street)
- Crosswalk placement
- Traffic signal location
- Number of signal heads
- Sequence of signal lights
- Traffic volume
- Parking restrictions
- Presence of parked vehicles
- School zone restrictions
- Length of pre-impact skid
- Visibility limitations from hedge

Figure 9–6 Impact area. *Courtesy of Colorado State Patrol.*

Figure 9–7 Post-impact evidence. *Courtesy of Colorado State Patrol.*

118 *Evaluation of the Highway Environment*

This accident occurred in an urban residential area on a two-way residential collector street. A car did not stop for the signal light and drove in front of the truck.

Data to Evaluate for Figures 9–8 and 9–9		
• Street identification	• Surface abnormalities	• Number of signal heads
• City and state location	• Curb and gutter compositions	• Sequence of signal lights
• Compass direction of street	• Curb and gutter style	• Traffic volume
• Vehicle travel directions	• Sidewalk presence	• Land use surrounding site
• Number of traffic lanes	• Sidewalk placement (both sides of street)	• Parking restrictions
• Widths of traffic lanes		• Presence of parked vehicles
• Overhead street light locations	• Pavement markings	• School zone restrictions
• Speed limit, speed advisories	• Condition of pavement markings	• Location of accident evidence
• Design of intersection	• Visibility of pavement markings	• Lane position—accident evidence
• Surface material and condition	• Regulatory signs	• Dimension of accident evidence
	• Warning-caution signs	
• Surface grade and cross slope	• Crosswalk placement	• Length of pre-impact skid
	• Traffic signal location	• Visibility limitations from hedge

Figure 9–8 View of truck approaching intersection.

Figure 9-9   View of impact area.

Your examination of the accident scene should also include an analysis and documentation of the impact area. In many situations, you will hear the phrase, "point of impact" being used. In reality, there is no *point* that establishes where impact occurred, as we are dealing with a variety of large vehicles colliding with each other and with other objects. If we consider passenger cars, sport-utility vehicles, and pick-up trucks, they typically have a width of approximately 5–6.5 feet. Commercial motor vehicles have a width of approximately 8–8.5 feet. Even if we consider a collision scrub tire mark, which was created by a tire of one vehicle, that mark may be up to several feet wide and long, depending on the impact configuration between the vehicles. Figure 9-10 shows the collision scrub tire marks, gouges, and liquid debris stains created during a head-on collision between a truck tractor and a pick-up truck.

So what point would you select for the location of impact? The evidence is clear that, in fact, there is an *area* of impact that should be documented, not just a single point. However, in almost every traffic accident report, the "point of impact" (also known as the approximate point of impact or APOI) will be measured by the officer. When you document the area on the pavement where a moving vehicle collided with something else, be cautious in *what* you identify as the impact area and *how* you identify it. You may consider using the phrase, "area of impact," instead of "point of impact."

120   *Evaluation of the Highway Environment*

Figure 9–10 Collision scrub tire marks, gouges, and liquid debris stains from head-on collision between truck tractor and pick-up truck create an area of impact.

The information you develop with respect to the dimensions of the vehicles involved will be utilized in this phase of your accident investigation. You may also need to consider the amount of front overhang, which is the longitudinal distance between the centerline of the front axle tire to the front bumper. This distance is depicted in Figure 9–11. Therefore, the point of first contact between the front of one vehicle and the rear or side of another may actually be several feet ahead of or behind the collision scrub tire mark created by the striking vehicle.

An examination of the impact area of different collisions indicates that each crash scene will have to be studied and evaluated to ensure that the proper impact configuration between the two vehicles is determined.

The arrow in Figure 9–12 identifies the area where the left front tire of the striking vehicle was positioned when the front bumper rear-ended the vehicle in front of it. Both vehicles begin to rotate counterclockwise, eventually creating arcing, post-impact, tire scuff marks (in the middle of the photograph). The straight skid mark on the left side of the photo was never attributed to a specific vehicle.

In the head-on crash between a car and a truck tractor in Figure 9–13, the area of impact is defined by the grouping of gouges in the pavement created by the undercarriage of the car during maximum engagement. The car was deflected to the right. The truck unit traveled toward the area shown in the upper-left corner of the photo, eventually entering the grass area adjacent to the road.

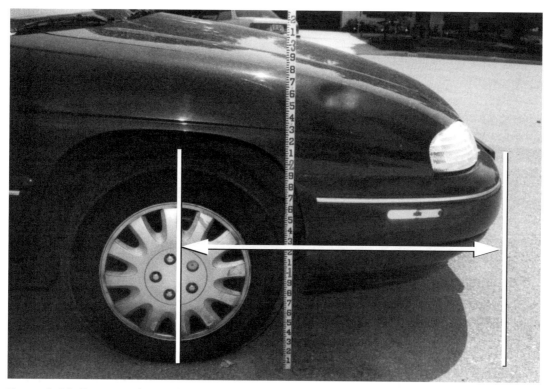

Figure 9–11 Front overhang (longitudinal distance between the centerline of the front axle tire to the front bumper) may need to be considered.

Figure 9–12 The arrow identifies the area where the left-front tire of the striking vehicle was positioned when the front bumper rear-ended the vehicle in front of it.

Figure 9-13 Area of impact is defined by the grouping of gouges in the pavement created by the undercarriage of the car during maximum engagement.

In this "side-swipe opposite-direction" collision between a car and a tractor-trailer combination, as seen in Figure 9-14, the area of impact was within the large debris field. Evidence of the impact was underneath the debris and determined by gouges in the pavement created by the undercarriage of the car. That evidence was in the area shown by the arrow.

Figure 9-14 Evidence of the impact was underneath the debris and determined by shallow gouges in the pavement. *Photo courtesy of Jerry Woods.*

Tire Mark Evaluation

A thorough knowledge of tire marks is one the important elements in understanding how vehicles traveled before, during, and after a crash. The charts in Figures 9–15 and 9–16 will assist you in identifying different types of tire marks created at different points in time. The sixteen boxes that follow offer examples of what those tire marks may look like when you view the highway. The boxes include information that will explain and identify the various types of tire marks.

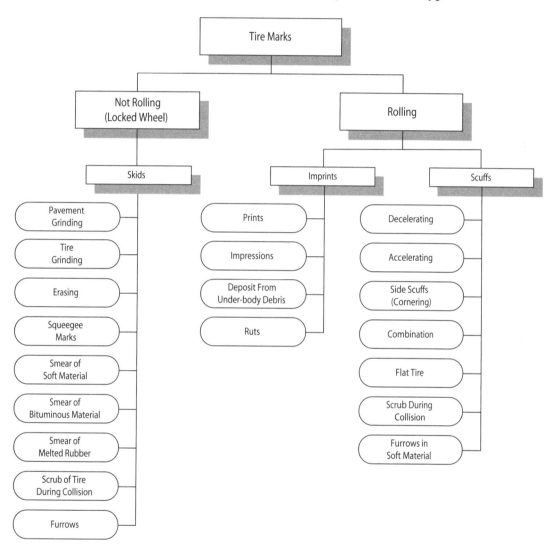

Figure 9–15 Tire mark flowchart.

There are two major categories of tire marks: those created by a wheel that is "rolling" and those created by a wheel that is "not rolling" (see Figure 9–16) If we now focus on the locked-wheel tire marks, or those tire marks created by wheels which are not rolling, we can examine the individual categories of tire marks within this classification.

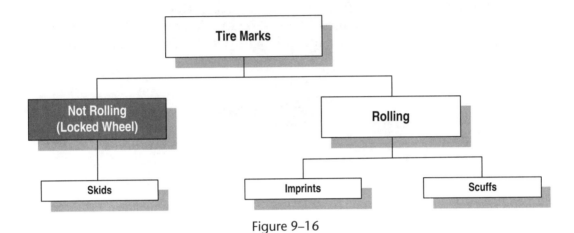

Figure 9–16

Locked Wheel Tire Marks—Pavement Grinding

- Skids
  - **Pavement Grinding**
  - Tire Grinding
  - Erasing
  - Squeegee Marks
  - Smear of Soft Material
  - Smear of Bituminous Material
  - Smear of Melted Rubber
  - Scrub of Tire During Collision
  - Furrows

Scratches from material on the surface, such as sand and gravel, or from studded snow tires.

Photos courtesy of Colorado State Patrol.

Locked Wheel Tire Marks—Tire Grinding

- Skids
  - Pavement Grinding
  - **Tire Grinding**
  - Erasing
  - Squeegee Marks
  - Smear of Soft Material
  - Smear of Bituminous Material
  - Smear of Melted Rubber
  - Scrub of Tire During Collision
  - Furrows

Tiny particles of tire material ground into the surface of the pavement.

Locked Wheel Tire Marks—Erasing

"Shadow" caused by sliding tire displacing material on surface.

Locked Wheel Tire Marks—Squeegee Marks

- Skids
  - Pavement Grinding
  - Tire Grinding
  - Erasing
  - **Squeegee Marks**
  - Smear of Soft Material
  - Smear of Bituminous Material
  - Smear of Melted Rubber
  - Scrub of Tire During Collision
  - Furrows

"Dry" area on a wet surface.

Similar to erasing.

Locked Wheel Tire Marks—Smear of Soft Material

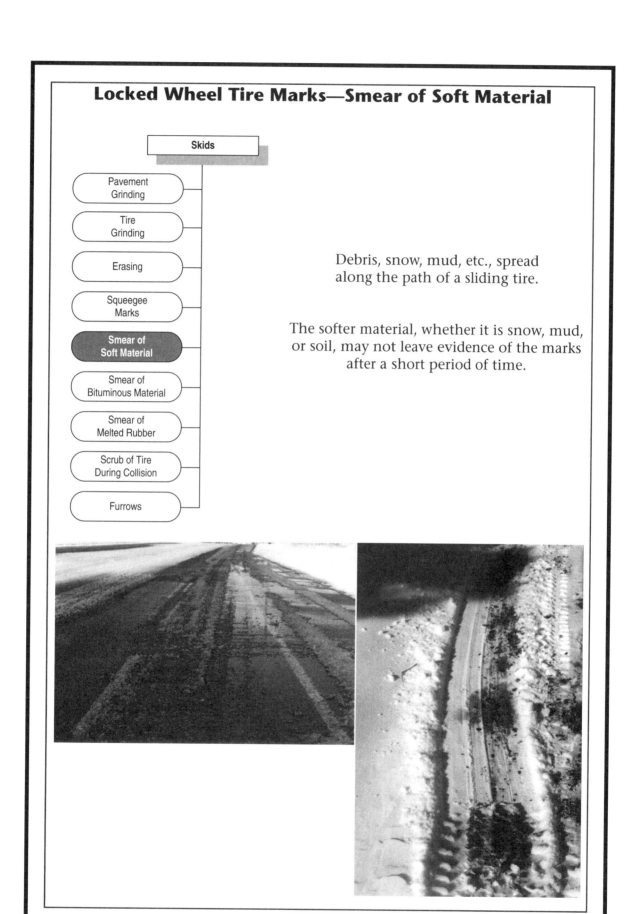

- Skids
  - Pavement Grinding
  - Tire Grinding
  - Erasing
  - Squeegee Marks
  - **Smear of Soft Material**
  - Smear of Bituminous Material
  - Smear of Melted Rubber
  - Scrub of Tire During Collision
  - Furrows

Debris, snow, mud, etc., spread along the path of a sliding tire.

The softer material, whether it is snow, mud, or soil, may not leave evidence of the marks after a short period of time.

Locked Wheel Tire Marks—Smear of Bituminous Material

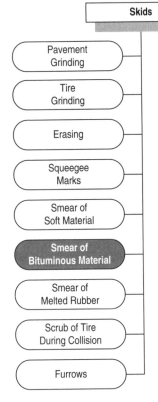

Asphalt or tar.

- Very dark (lasts a long time)
- Tire hot enough to melt rubber
- Typically on roads with large amounts of oil in asphalt mix

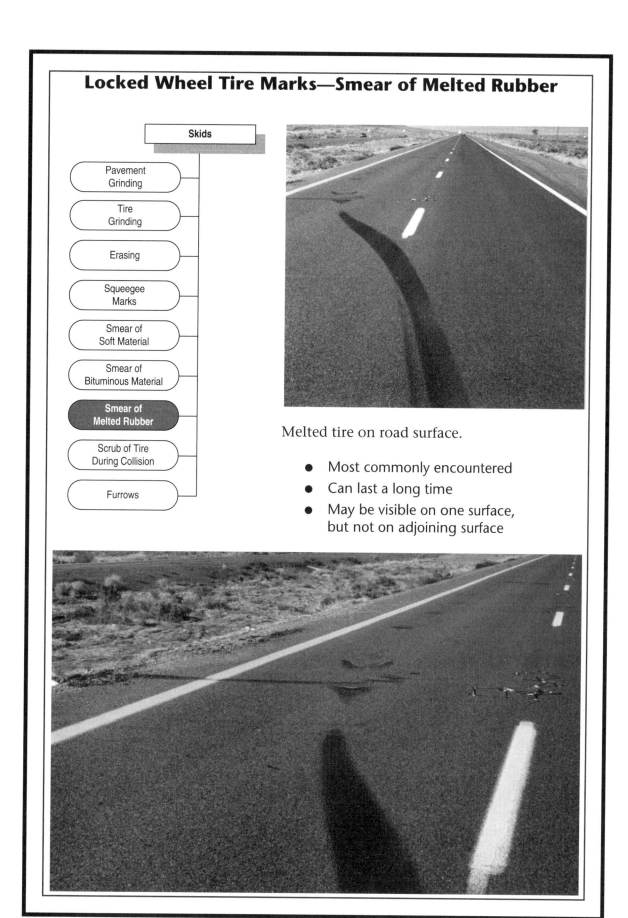

Locked Wheel Tire Marks—Scrub of Tire During Collision

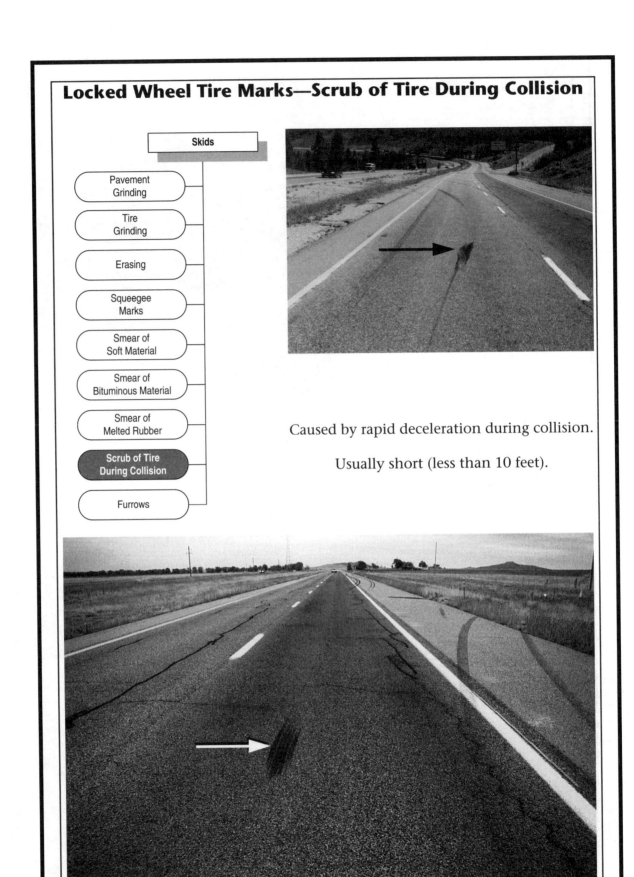

Caused by rapid deceleration during collision.

Usually short (less than 10 feet).

Locked Wheel Tire Marks—Furrows

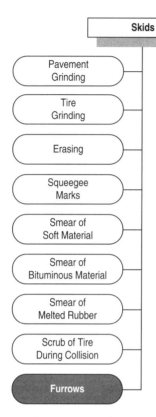

The trench left when skidding on a soft surface.

- Sliding tire pushes loose material ahead and to the side
- Bottom of furrow is roughened and irregular

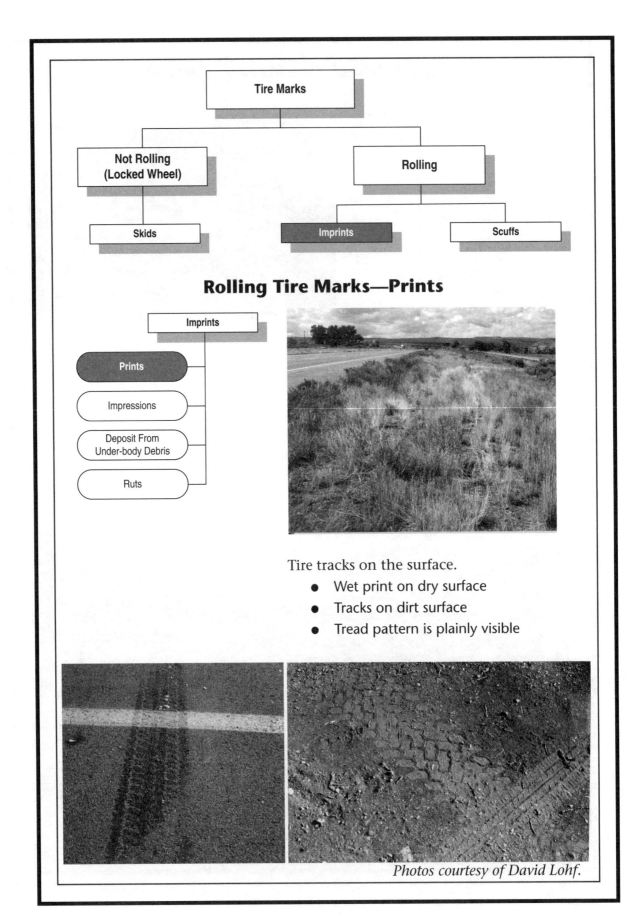

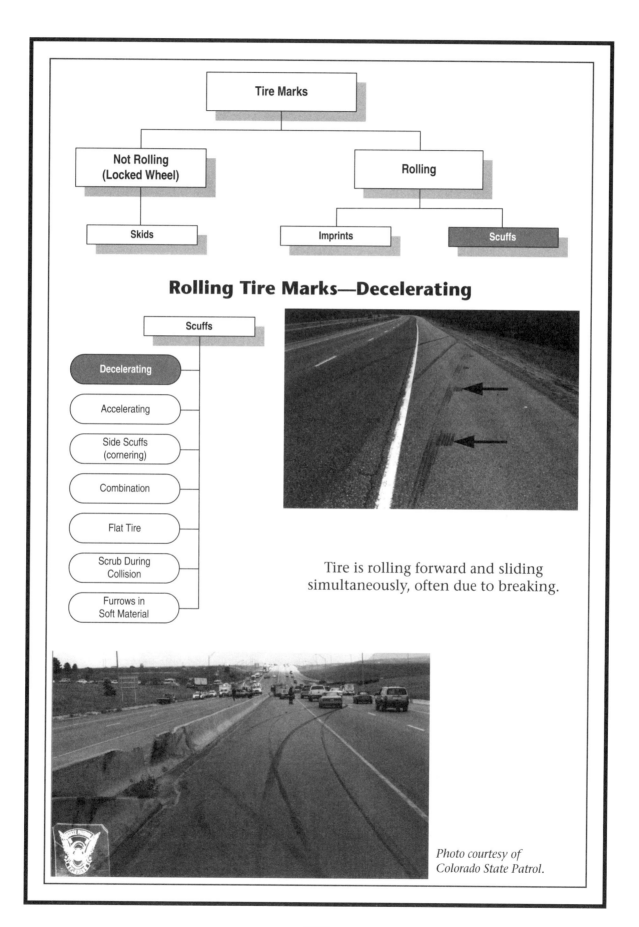

Photo courtesy of Colorado State Patrol.

Rolling Tire Marks—Accelerating

- Scuffs
  - Decelerating
  - **Accelerating**
  - Side Scuffs (cornering)
  - Combination
  - Flat Tire
  - Scrub During Collision
  - Furrows in Soft Material

Photos courtesy of David Lohf.

Caused by sudden application of power to drive wheels.

Rolling Tire Marks—Side Scuffs (cornering)

- Scuffs
  - Decelerating
  - Accelerating
  - **Side Scuffs (cornering)**
  - Combination
  - Flat Tire
  - Scrub During Collision
  - Furrows in Soft Material

Tire is rotating and sliding sideways, creating diagonal striation marks.

Rolling Tire Marks—Combination

- Scuffs
  - Decelerating
  - Accelerating
  - Side Scuffs (cornering)
  - **Combination**
  - Flat Tire
  - Scrub During Collision
  - Furrows in Soft Material

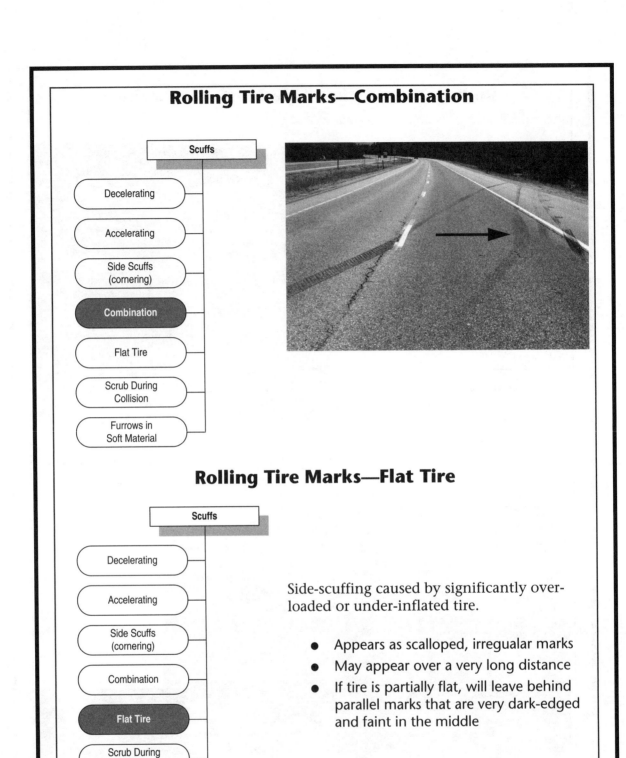

Rolling Tire Marks—Flat Tire

- Scuffs
  - Decelerating
  - Accelerating
  - Side Scuffs (cornering)
  - Combination
  - **Flat Tire**
  - Scrub During Collision
  - Furrows in Soft Material

Side-scuffing caused by significantly over-loaded or under-inflated tire.

- Appears as scalloped, irregular marks
- May appear over a very long distance
- If tire is partially flat, will leave behind parallel marks that are very dark-edged and faint in the middle

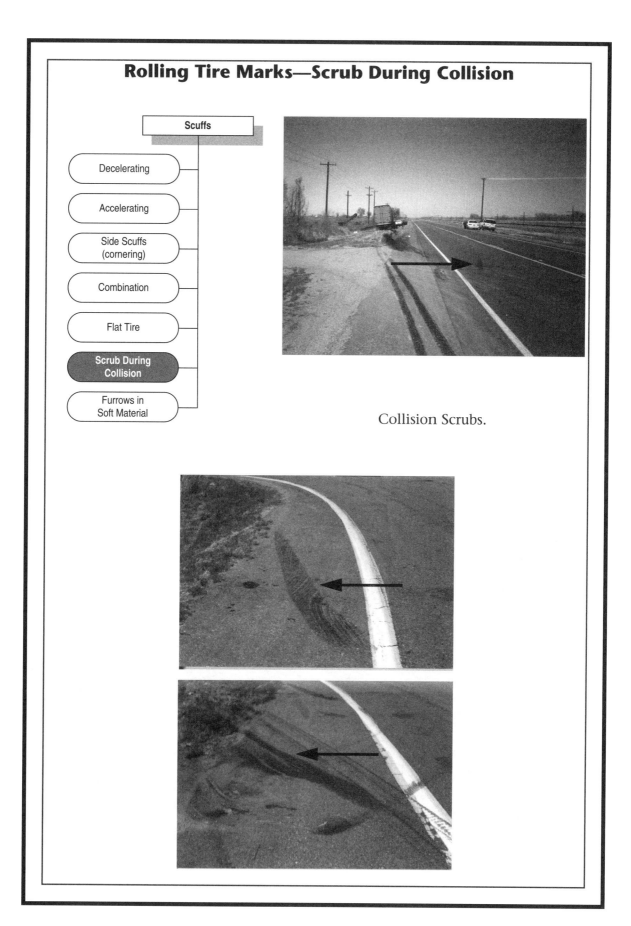

Collision Scrubs.

Rolling Tire Marks—Furrows in Soft Material

- Scuffs
  - Decelerating
  - Accelerating
  - Side Scuffs (cornering)
  - Combination
  - Flat Tire
  - Scrub During Collision
  - **Furrows in Soft Material**

10
Interviewing

Interviewing can be an effective way to gain knowledge about a traffic accident event. An interview will allow you to develop relevant information regarding the collision event and to gain knowledge about all three elements of the traffic crash—the human, the vehicle, and the environmental elements. Understand that any information you develop from an interview will probably be subjective information regarding these three elements, as opposed to objective information. The reason for this is the human perspective. We all interpret what we recall, what we understand, and what we perceive about the traffic accident.

Interviewing is a skill at which many people are not very proficient. They are nervous about questioning someone else, they are uncertain how to handle the suspicion and frustration that may result from questioning, they are uncertain of their abilities, and often, they anticipate rejection from the person they want to interview. One of the important things to remember is that if you approach this task as a professional and as an honest, caring individual, you will often succeed with the interview.

Who are you going to interview? The answer is anyone who might have some relevant information useful to your investigation of the traffic accident. Recall our previous review of the human element of the accident investigation. Your interviews might include:

- Drivers involved in the event and any passengers that may have occupied any of the vehicles involved in the event.
- Independent witnesses to the crash event or its aftermath. These witnesses may be passengers in a vehicle, other motorists or vehicle passengers in the immediate area of the crash, motorists who saw the driving behaviors of the vehicle operators before the crash event, or bystanders who may have been walking or standing near the crash location.
- People who have some relationship to the accident event. These people would include police officers, fire and emergency response personnel, ambulance and paramedic responders, tow truck drivers, HAZMAT clean-up responders, independent claims adjusters, independent photographers, and news media personnel who responded to the accident scene.

All of these people may offer information on some aspect of the activities or events that occurred before, during, and after the accident, as well as the conditions and circumstances surrounding the accident. The only limitations are yours, especially if you don't inquire into all potential sources of information.

Whether you are conducting an informal discussion at the scene of the traffic accident or a formal interview, such as a post-accident interview of your driver in the conference room of your corporate offices, experts in the interview process offer two keys to success: *prepare* and *listen*.

Keys to Interview Success

Prepare for your interview. The preparation portion of the interview allows you to think about what you want to achieve by interviewing a person. Think about how you are going to ask the questions. Think about what you will be asking. Remember that conducting an interview and/or obtaining a statement is not a goal of accident investigation, but a method of collecting information.

As you prepare for the interview, keep in mind some basic principles of obtaining information:

- Be objective in your questioning
- Don't reveal your personal feelings; remain strictly professional and courteous
- Be positive and professional in your approach to the witnesses
- Be specific. Ask clear questions and make sure each person understands the question
- Don't suggest answers by the way in which you word your questions
- Avoid conflict in your questions
- Be adaptable during your interview and adjust your questions to the comfort level of the person you are interviewing
- Verify statements
  - Check earlier or other statements made by this person

- Check against physical evidence
- Check against statements of other drivers and/or witnesses
- Consider when and where to question
  - Select the time, place, and method of questioning
  - Keep questions flexible
  - Question privately to avoid conflict and distraction
  - Avoid an informal atmosphere with many people listening
- Be aware of two kinds of information you will be seeking:
  - Routine—license, registration, report form
  - Factual—how and why the accident occurred

You may prefer to write out questions that you will be asking. You may want to utilize a prepared, but generalized, question format. Some interviewers are comfortable working off an outline or checklist. You need to prepare yourself mentally for what and how you will interview the person.

The second key to success in interviewing is to *listen*. Listen to what the person is telling you. This seems so simple, but many people overlook this basic activity. This is especially important if you are conducting the interview by telephone, as you don't have the advantage of viewing the person's facial expressions, mannerisms, or other physical responses to your questions. If you are not listening to what the person is telling you (maybe you are thinking about your next question), you may miss something important. The person may offer a hint about a subject that you will miss if you are not listening. You will lose the opportunity to inquire about that hint. And if the person you are interviewing gets the impression that you are not interested in *listening* to them, he or she might cut the interview short because of your inferred attitude.

What do you ask a person regarding an accident? The best way to answer that question is to focus on "who, what, when, where, how, and why" questions. The following are some examples of those types of questions that deal with the driver's trip before the accident occurred:

- Who determined what route you took?
- Where were you going?
- What route were you intending to take?
- How many times have you traveled on this highway?
- When were you supposed to be at your destination?
- Why did you choose to travel on this highway?
- Have you previously traveled on this road at the same time of day as the accident?

As you can see, it is easy to develop questions on one small aspect of the accident by utilizing a generalized approach. Each interview may have similar areas of inquiry. However, each interview needs to be tailored to the individual being questioned, their perspective regarding the interview, and the circumstances of the interview.

Possible Areas of Inquiry During Post-Accident Interviews

The following outline highlights areas you may want to include in your interview.

I. IDENTIFICATION DATA
    A. Full name
    B. Street address, city, state, zip code
    C. Telephone numbers: home, business, cell
    D. Drivers license number and state, restrictions on license
    E. Vehicle identification
        1. Year and make
        2. Complete model information
        3. Vehicle body style
        4. Color
        5. Engine and transmission information
    F. Number of occupants, where each was seated, if seat belts used
    G. Driving experience in years, type of driving experience
    H. Annual estimated mileage
    I. Any previous driver education: what, where, when
    J. Vehicle familiarity
        1. Years
        2. Mileage
    K. Occupation, if driving-related
    L. Date of birth
    M. Marital status

II. PRE-CRASH DATA
    A. Trip plan
        1. Origin of trip
            a. Time started driving
            b. Location of start of trip

2. Destination
   a. Location of destination
   b. Need to arrive at specific time or time constraints
3. Purpose of trip
4. Familiarity with route and area

III. ACTIVITIES PRIOR TO COLLISION
   A. Physical condition
      1. Sleep amount prior to accident
      2. Work activities prior to accident
      3. Recreational activities prior to accident
      4. Distance of travel on this trip
   B. Immediate condition and/or infirmities
   C. Any pressures or stress while driving, state of mind
   D. Any use of alcoholic beverages, illegal drugs, or prescription drugs
   E. Smoking or eating during trip or just before collision

IV. DESCRIPTION OF COLLISION
   A. Exact details regarding the vehicle's travel before impact
   B. Direction of travel and identification of street/road/highway
   C. Lane of travel, any lane changes just before accident
   D. Estimated speed of the vehicle and how that speed determined
   E. Traffic conditions in all directions
   F. Point of first awareness of danger (try to relate to a location, a distance, or a specific point such as a driveway, sign, curve in the road, or a change in lane configuration)
   G. Decision/reaction to hazard
   H. Observations of driver
      1. Activities of other vehicles
      2. Any traffic control signals controlling vehicle movement
   I. View obstructions or view interferences
   J. Distractions
      1. Any inside the vehicle
      2. Any outside the vehicle
         a. Construction area
         b. Heavy traffic or traffic congestion
         c. Other accident or activity near the highway
   K. Any interior areas of vehicle struck during the collision
   L. Anyone using cell phone just prior to collision

M. Ejection from the vehicle
N. Anyone struck by any loose objects inside the vehicle
O. Location of interviewee when the crash was viewed

V. POST-COLLISION DATA
  A. Description of post-crash travel
  B. Driver actions
  C. Injuries sustained
  D. Exit from the vehicle by driver and passengers; if exit assisted
  E. Source rendering first aid
  F. Ambulance service; time to arrive at scene after accident
  G. Manner of leaving accident scene
    1. Able to drive away
    2. Other drivers or passengers walking around the scene

VI. VEHICLE REVIEW
  A. Classification of vehicle condition; any mechanical problems noted
  B. Any equipment malfunction claimed that is related to collision
  C. Vehicle last serviced/repaired: who, when, where, what done
  D. Luggage or cargo inside the vehicle
    1. What it was and approximate weight
    2. Location of that cargo
  E. Exact location of vehicle damage resulting from the collision
  F. Any preexisting or unrelated damage on vehicle
  G. If vehicle drivable after the collision
  H. Restraint system used
    1. Seat belt
    2. Shoulder harness

VII. COLLISION REVIEW
  A. Approach path of all vehicles (lane position, turning position)
  B. Location of the area of impact
    1. On each vehicle
    2. On the road
  C. Speed at impact of each vehicle
  D. Time of impact; if sunlight, sunset, sunrise conditions a factor
  E. Driver and passenger action at impact
  F. Final resting positions of all vehicles
  G. Description of injuries sustained by all vehicle occupants

    H. Weather and road surface conditions

VIII. OTHER AREAS OF INQUIRY
    A. If accident preventable or avoidable by either driver
    B. Reasonable action by others
        1. Evasive actions to avoid the collision
        2. Timeliness of evasive actions
    C. Any experience in a similar situation or attempts to avoid an accident
    D. Highway design or condition; any contribution to accident
    E. Speed limit and the speed of other traffic before accident
    F. Highway maintenance problems
    G. Traffic control devices in accident area
        1. Signs
            a. Type and location
            b. Visibility
        2. Signals
            a. Location
            b. Visibility
            c. Sequence/operation
        3. Pavement markings
            a. Type and location
            b. Visibility
    H. Preferred lane of travel
        1. This road
        2. Generally
    I. Posted speed limit
    J. Safe following distance at indicated speed
    K. Previous accidents
        1. How many
        2. Most recent
        3. Involvement in similar type of accident

A

Resources

Organizations and Databases

Accident Reconstruction Network (www.accidentreconstruction.com)

> A resource for a variety of supplies, research, resources, and information for the traffic accident investigator.

ACTAR-Accreditation Commission for Traffic Accident Reconstruction (www.actar.org)

> A resource for a variety of topics, including:
> - Traffic accident reports and overlays for all 50 states
> - Links to traffic crash research and CMV-related sites
> - Directory of accredited reconstructionists

Bridgestone-Firestone Commercial Tires—www.trucktires.com

> A publication and resource center about commercial motor vehicle tires.

NCDC-NOAA-NWS website—www.ncdc.noaa.gov

> A resource for obtaining historical National Weather Service data for a given area.

Traffic Accident Reconstruction Origin (TARO) website (www.tarorigin.com)

> A resource for accident research, supplies, resources, and information relating to accident investigation.

TerraServer-USA (www.terraserver-usa.com)

A free resource for aerial photographs and topographical maps of the United States created by the U.S. Geological Survey.

U.S. Naval Observatory (www.usno.navy.mil)

A resource for obtaining sunrise, sunset, moonrise, and moonset times.

USDOT-FMCSA Site (www.fmcsa.dot.gov)

A resource for information for CMV operations.

Vehicle Dimensions (www.rec-tec.com)

A free downloadable database of vehicle dimensions, provided by Transport Canada.

Classes

Classes teaching the fundamentals of traffic accident investigation and traffic accident reconstruction are available from these universities throughout the United States.

University of North Florida Institute of Police Technology and Management, Jacksonville, Florida, 904-620-4786.

North American Transportation Management Institute, 720-887-0835

Northwestern University Center for Public Safety, Evanston, Illinois (formerly known as The Traffic Institute), 800-323-4011.

University of California Riverside Extension, 909-787-5804.

Publications

Daily, John and Nathan Shigemura (1997), *Fundamentals of Applied Physics for Traffic Accident Investigators,* Institute of Police Technology and Management.

Daily, John (1988), *Fundamentals of Traffic Accident Reconstruction,* Institute of Police Technology and Management.

Lofgren, M.J. (1981), *Handbook for the Accident Reconstructionist,* Arnold's Printing.

Baker, J. Stannard (1975), *Traffic Accident Investigation Manual,* Northwestern University Traffic Institute.

Fricke, Lynn B. (1990), *Traffic Accident Reconstruction,* Northwestern University Traffic Institute.

Manual on Uniform Traffic Control Devices, A federal government publication providing information on traffic signs, pavement markings and traffic signals.

Index

A

accident 1–2
    federal definitions 4–8
    general definitions 2–3
    law enforcement investigation 26–30
    legal definition 26
    reports 30–34
    scene 35–38
    state definitions 3–4
    time frames 41–43
accident investigation
    classes 150
    equipment and supplies 19–24
    factors 39–52
        environment element 41
        human element 40
        time frames 41–43
        traffic crash elements 43–52
        vehicle element 40
    highway environment 109–122
    interviews 141–147
    law enforcement 26–30
    organizations and databases 149–150
    photographic techniques 79–90
    planning and preparation 9–18
    publications 150
    recording techniques 55–78
    traffic accident reports 30–34
    vehicle information 93–108
    websites 149–150
    working with police 35–38
Accident Reconstruction Network 149
accident templates 76–78

Accreditation Commission for Traffic Accident Reconstruction (ACTAR) 15, 149
advisory bodies 31
American Association of Motor Vehicle Administrators 31
anticipation 45
anti-lock braking system (ABS) 47
approximate point of impact (APOI) 119
atmospheric conditions 112
attorneys 17

B

bituminous material, smear of 130
body repair facilities 17
body style 94

C

camera 20
claims adjusters 15
classes 150
Code of Federal Regulations 4
cognition 45
collision 1
    activities prior to 145
    description 145–146
    disengagement 49
    first harmful event 48–49
    initial contact 49

collision (continued)
    maximum engagement 49
    post-collision data 146
    review 146
    rolling tire marks 139
Colorado Revised Statutes 3
commercial motor carriers
    in-house resources 12–14
    outside resources 14–18
    planning and preparation 9–18
    policies and procedures 10–12
contact damage 101
contrast 45
controlled rest position 52
coordinate measuring system 62–67
corporate resources (in-house) 12–14
    administration department 12
    dispatch department 13
    driver 14
    human resources department 12
    legal department 13
    maintenance department 12
    safety department 14
    See also outside resources
counter report 28
crash 1
    elements 43–52
    encroachment 45–46
    evasive action 46–47
    final rest position 52
    first harmful event 48–49
    initial contact 49
    maximum engagement 49
    perception, reaction, and response 44–45
    possible perception 43–44
    separation 49
    time frames 41–43

D

damage 5
    categories 101
    disabling 5
    documentation 96–100
    evaluation 101–108
debris 62
disengagement of vehicles 49
dispatch department 13, 14

Drive 94
drive train 94
drivers 14, 142

E

eccentricity 45
encroachment 45–46
engines 94
environment element (accidents) 41
equipment and supplies 19–24
    measuring tools 20
    photographic equipment 20
    recording devices 20
    reference resources 21
    safety and protective devices 21
    storage and transportation 22–23
erasing (tire marks) 127
evasive action 46–47
evidence
    life span of 110–111
    sketching 62

F

fatality 5
Federal Motor Carrier Safety Regulations 4–8
field sketching 56–62
    items 58–62
    materials 58
final rest position 52
first harmful event 48–49
flat tire 138
furrows (tire marks) 133, 140

G

general rules 11

H

hazardous materials cleanup and remediation 16

highway 6–7
   atmospheric conditions 112
   information 110–122
   photographs 87–90
   sketching 59–60
   surface conditions 113–114
   visibility limitations 111–113
human element (accidents) 40

I

induced damage 101
in-house resources 12–14
   administration department 12
   dispatch department 13
   driver 14
   human resources department 12
   legal department 13
   maintenance department 12
   safety department 14
initial contact 49
insurance claims adjusters 15
interviews 141–147
   areas of inquiry 144–147
      activities prior to collision 145
      collision review 146
      description of collision 145–146
      identification data 144
      post-collision data 146
      pre-crash data 144
      vehicle review 146
   listening in 143
   questions 143
   techniques 142–144
   See also accident investigation

L

law enforcement investigation 26–30
leased equipment 18
legal department 13
legal representatives 17
load clean-up services 15
locked wheel tire marks
   erasing 127
   furrows 133
   pavement grinding 125
   scrub during collision 132
   smear of bituminous material 130
   smear of melted rubber 131
   smear of soft material 129
   squeegee marks 128
   tire grinding 126
   See also rolling tire marks

M

maintenance and repair facilities 17
Manual on Classification of Motor Vehicle Traffic Accidents 2
marking materials 72
maximum engagement 49
measuring tools 20
medical facilities 18
melted rubber, smear of 131
motor carriers
   in-house resources 12–14
   outside resources 14–18
   planning and preparation 9–18
   policies and procedures 10–12
motor vehicle accidents. See traffic accidents

N

National Highway Traffic Safety Administration 39
negligence 30

O

outside resources 14–18
   hazardous materials clean-up/ remediation 16
   insurance claims adjusters 15
   leased equipment 18
   legal representatives 17
   load clean-up services 15
   maintenance and repair facilities 17
   medical facilities 18
   owner-operators 18
   towing services 16
   traffic accident reconstructionist 15

outside resources (continued)
    See also corporate resources (in-house)
owner-operators 18

P

pavement grinding (tire marks) 125
pavement markings 52
perception 44–45
photographic equipment 20
photographic techniques 79–90
    highway photographs 87–90
    vehicle examinations 79–85
point of impact 119
policies and procedures 10–12
possible perception 43–44
protective devices 21
publications 150

R

reaction 44–45
recording devices 20
recording techniques
    accident templates 76–78
    coordinate measuring system 62–67
    field sketching 56–62
    triangulation measurements 70–74
reference lines 63
reference point 63–67
reference resources 21
remediation services 16
repair facilities 17
response 44–45
response complexity 45
restraint system 146
road design, sketching 60–61
rolling tire marks
    accelerating 136
    combination 138
    decelerating 135
    flat tire 138
    furrows 140
    prints 134
    scrub during collision 139
    side scuffs 137
    See also locked wheel tire marks
rubber, melted 131

S

safety and protective devices 21
safety department 14
seat belts 146
separation of vehicles 49
shoulder harness 146
side scuffs (tire marks) 137
signals 147
signs 147
specific rules 11
squeegee marks (tire marks) 128
statutes 3–4
stimulus strength 45
surface conditions (highway) 113–114

T

TerraServer-USA 150
time frames 41–43
tire grinding 126
tire marks 123–124
    accelerating 136
    combination 138
    decelerating 135
    erasing 127
    flat tire 138
    furrows 133, 140
    pavement grinding 125
    prints 134
    scrub during collision 132, 139
    side scuffs 137
    sketching 61–62
    smear of bituminous material 130
    smear of melted rubber 131
    smear of soft material 129
    squeegee marks 128
    tire grinding 126
towing services 16
Traffic Accident Investigation Manual 101
Traffic Accident Reconstruction Origin
    (TARO) 149
traffic accident reconstructionist 15
traffic accidents
    environment element 41
    federal definitions 4–8
    general definitions 2–3
    highway environment evaluation
        116–119

    human element 40
    law enforcement investigation 26–30
    legal definition 26
    recording techniques 55–78
    reports 30–34
    scene 35–38
    state definitions 3–4
    time frames 41–43
    vehicle element 40
traffic control devices 147
traffic control devices, sketching 60
triangulation measurement system 70–74
trip plan 144

U

uncontrolled rest position 52
Uniform Traffic Code 26
Uniform Vehicle Code 3

V

vehicle element (accidents) 40
vehicle examinations 79–85
vehicle identification number (VIN) 95
vehicle information 93–108
    components and safety items 98–99
    damage 101–108
    documentation 94–95
    evidences of previous damages 96–100
    photographic documentation 79–85, 99
    sources 100
vehicle review 146
view interference 112
view obstruction 111
visibility limitations 111–113

W

websites 149–150
witnesses 142

HV 8079.55 .W49 2005

Wheat, Arnold G.

Accident investigation
training manual

DATE DUE

SCHENECTADY COUNTY
COMMUNITY COLLEGE
LIBRARY